A Practical Poultry Plant For Southern California

by E. Pryce Mitchell

with an introduction by Jackson Chambers

Self Reliance Books

Get more historic titles on animal and stock breeding, gardening and old fashioned skills by visiting us at:

http://selfreliancebooks.blogspot.com/

Introduction

I am pleased to present yet another title on Poultry.

The work is in the Public Domain and is re-printed here in accordance with Federal Laws.

As with all reprinted books of this age that are intended to perfectly reproduce the original edition, considerable pains and effort had to be undertaken to correct fading and sometimes outright damage to existing proofs of this title. At times, this task is quite monumental, requiring an almost total "rebuilding" of some pages from digital proofs of multiple copies. Despite this, imperfections still sometimes exist in the final proof and may detract from the visual appearance of the text.

I hope you enjoy reading this book as much as I enjoyed making it available to readers again.

Jackson Chambers

DEDICATED

TO

"DOUBTING THOMAS"

PREFACE

At sea a man is considered to have no time absolutely his own, but is liable to be called on at any moment; he expects this before he ships and is therefore prepared for it.

This same rule applies to the poultry business, so unless one is prepared to give up a good deal of his "watch below," Sundays included, it were better left alone, for if he gets tired of it in a year or so, much of the investment is likely to be lost when selling out.

In case a reader of this book intends to take it up seriously, and requires more information on certain points, the writer will be glad to hear from him.

E. PRYCE MITCHELL.

Santa Barbara, California.

CONTENTS

CHAPTER I.

CONTENTS

CHAPTER XIII.

CHAPTER XIV.

CHAPTER XV.

CHAPTER XVI.

CHAPTER XVII.

COCKERELS READY FOR CULLING

A PRACTICAL POULTRY PLANT

CHAPTER I.

INTRODUCTORY

Five years ago I commenced raising chickens for the market; my only experience was some theories that I had worked out during some twenty odd years at sea. I had never seen a poultry plant, and felt rather cheered up by hearing how several others had failed in that particular line of business; with the courage of ignorance I started in, trusting in hard work and common sense to carry me through.

The first year I raised "soft roasters," and found a ready market for them at fifty cents each; people came and bought all I could raise, and praised them to my heart's content, as long as I kept the price at fifty cents.

On going over my accounts at the end of the year I found that those soft roasters had cost me from fifty-five to sixty cents each to raise, and that I had lost $300 cash on the year's work; but I had found lots of experience.

The second year I turned everything into Barred Plymouth Rocks, keeping the pullets for layers and caponizing the cockerels. I kept the pullets in flocks of sixty, and at the end of the year I had cleared $250, over the expenses, and added to my experience.

The third year I invested in a good many settings from the finest stock in the country, at five dollars a setting, and added Light Brahmas and White Leghorns to my flocks, making three breeds in all. I caponized over three hundred birds, and sold nearly all the eggs as market eggs. I netted about $500 that year, and began to get my bearings, for I learned that the smaller breeds, such as Leghorns, can be kept in flocks of from fifty to a hundred and lay well, while the heavy birds, like the Brahmas, have to be kept in small flocks in order to get the best results.

Up to this time I had been using only an acre and a half of ground, but I now felt safe in adding two acres and a half more, making four in all. I also added another breed to my stock, the

Buff Orpington, and began the fourth year with several fine yards of Barred Plymouth Rocks, Light Brahmas and Buff Orpingtons; also, six hundred White Leghorn pullets, just beginning to lay.

I began to advertise settings of fifteen eggs at $2 per setting, and sold a fair amount; all the surplus eggs went to market. I also caponized over four hundred cockerels. At the end of the year I was over $700 to the good, and a new idea was beginning to dawn on me—I say "new," because it was new to me—that money could be made in buying, as well as in selling. Up to this time I had been buying my grain from month to month, as I needed it, I now began to learn that I could save from ten to twenty-five per cent by buying a year's supply at the right season, just after harvest.

The fifth year found me with some splendid stock birds and eight hundred Leghorn pullets and hens. I was now satisfied that market eggs were going to be the principal source of income, and that it paid to sell off all the young cockerels as soon as possible, instead of making capons of them, and to fill up their place with pullets.

I made a contract with a large hotel to take all my market eggs, and agreed to send them in every day, to insure their getting them strictly fresh. During the year the demand for settings increased steadily and I sold a great many to people in all parts of the state. My eggs hatched strong, healthy chickens, and the fertility ran high. I was careful to sell only from my best stock, and was always ready to duplicate free, in cases where the eggs had arrived in bad condition through careless handling during transit.

The fifth year is now at an end. My sales have been about $3000. The food has cost $1,200, and labor, $500; I am therefore $1,300 ahead by the year's work. Had I employed one boy instead of two, and done some of the manual work myself, I could have netted a clear $1,500. I have learned more this year than I would have thought possible, and still think I am going to learn something new each year about chickens, as long as I continue to raise them. I know more today than I did five years ago, but I am a long way from being an expert. I am simply a practical man, running a poultry plant on strictly business lines, under conditions suited to this climate, where poultry grows more slowly than in a country where there are fixed seasons, where feed is dearer than in most parts of the United States, and where the cost of labor is almost prohibitive.

This little account of my work makes no pretence of being a complete manual of poultry raising. There are plenty of excellent works of that kind already on the market, some of which I shall have occasion to mention later on. I am only trying to give a plain statement of facts about poultry raising in Southern California, where conditions are essentially different from those that prevail in the East. The knowledge that I have now would have saved me much time and money if I could have availed myself of it at the start; and it is with the hope of saving other beginners from similar errors that I am offering this record of my experience.

CHAPTER II.

GETTING BEARINGS

There is little danger of overdoing the poultry business in Southern California for many years to come. Tourists swarm here from all parts of the country almost all through the year; while eggs and poultry in cold storage are shipped in from other states by the carload (1850 car loads last year); every year sees new poultry plants started, which after a longer or shorter period—generally shorter—are abandoned as failures.

The reasons are many; lack of capital, lack of experience and push, inability to look ahead, getting tired too soon, are all frequent causes of short-lived experiments; but the principal reasons are ignorance of the markets, and plants that are laid out poorly and worked without system. Perhaps this last reason is most often the true one. Some people are naturally good organizers, and these can generally get good results without much outside help; but the majority will succeed better and quicker by following in the footsteps of one who has made a practical success of his business. Avoid the person who can tell you all about it, but has nothing to show but a plant on paper.

It is impossible to fix the exact capital necessary for starting the poultry business; some will say one thing, while others will think quite differently. In this book I am going to work out a plant that will take all of one man's time, and give him an income of about $1,500 a year, net. I would not advise a larger plant than one man can manage, by giving all his time to it.

The poultry business keeps one very close to it all the time, and the work becomes very monotonous. I consider it a safer plan to be content with a smaller income, and pay a boy to do the rough work. With a little supervision the average boy can be taught to do all the cleaning and to do it thoroughly. Too much routine work is apt to make one restless and impatient, and when that spirit manifests itself, the income is sure to suffer; so I strongly advocate keeping a good, steady boy to do all the cleaning and hard work, leaving the owner free, to a certain extent, to give his time and brains to general supervision, feeding, incubating, brooding and care of the young stock.

The orchard and kitchen garden will amply repay him for all the care he can give them. I mention kitchen garden and orchard, as I am considering in these pages a home, as well as a poultry business.

Let us take the case of a stranger arriving in Los Angeles with his family and, say, $10,000. Should he have less than this I would advise him to spend less on his house, but to be prepared to invest at least $5,000 in the business from which he expects to derive his income.

He will at first go for a few days to a quiet hotel or boarding house, where the charges are moderate, and then rent a small house or flat, furnished or unfurnished, according to his needs. Good, serviceable furniture can be bought at a moderate price, but it would be as well not to get more than is absolutely needed until the future home is located.

Two or three months can be profitably spent in visiting the different towns of Southern California, none of which are more than a few hours by train from Los Angeles. The pleasure and health resorts should receive the most attention, for there will be found the greatest demand for choice poultry and fresh eggs. Let the final decision as to a resting place be arrived at only after much thought and consideration. Remember that a month, or even two months, spent in weighing the pros and cons of the place that is to be the home for perhaps many years, is time profitably spent, and probably means the success of the new undertaking. There are beautiful places in Southern California, and deadly dreary places, and the higher prices commanded by land in attractive localities is money well invested. A point that requires much thought and personal experience, or, rather, the experience of those longer in the country, is the choice of land that has a reasonable prospect of rising in value in the near future.

Southern California is a young and growing country, and in many localities land values have increased by leaps and bounds within the last few years. For instance, I am living on land that was bought for $50 an acre twenty years ago. Thirteen years later it brought $500 an acre, and at the present writing, it and most of the surrounding lands are considered reasonable at $1,000 an acre. There is an encouraging possibility that after ten or fifteen years of work, the land that was bought by the acre can be sold by the lot, and enough be realized to provide a comfortable competency for one's declining years.

Until one has looked about for himself and found a place that appeals to him, it is wise to shun real estate agents and others who have land to sell. In fact, never let anyone know you are looking for a place, for if you do you will have a task before you that only one very familiar with the different localities and conditions can possibly bring to a successful issue, that is, the choice of the best place amongst the many that will be offered. Of course, one may strike it lucky, but it is much better to take the safer course, and decide only after careful investigation from different points, even if the living expenses do go on a little longer. Nothing is more common and nothing more disheartening than to find after a year or two that one has made the wrong selection in the choice of a home. It is done every day here.

One of the first things that strikes the stranger, is the feeling of jealousy between the different towns. It is a rare thing to hear a good word spoken of "the other town." Personal observation is therefore the only safe way, and both experience and observation lead me to lay much stress on this point in advising a beginner.

I must give one other warning. BEWARE OF THE EN-THUSIAST! He has nothing to sell, nothing to gain, and is therefore the more dangerous. The average man, hearing a would-be seller dilate on the advantages of his piece of property, naturally allows his common sense to suggest that there are two sides to a bargain. Unfortunately, this common sense is often absent or asleep when the enlivening views of the enthusiast are being set forth, especially if they happen to be about a piece of property that the listener happens to fancy. The enthusiast is usually a well-meaning person, but almost always unreliable as a counsellor, for to get a plain, unvarnished, conservative opinion from him is an utter impossibility. As a guide he is worse than useless. Mind, I don't believe in letting the pessimist have all the hearing; the best way is to try to strike a happy medium by weighing both sides, and then using your careful judgment.

I fear my reader will think I am a good deal off my course, but I am trying to fill the needs that I myself experienced as a stranger here, some six years ago. A good start is as important in this business as in any other; and the man who wants to make a living from poultry can not afford to make avoidable mistakes.

In the preceding pages I advised keeping the plant within the compass of one man's labor. This is sure to strike the ambitious beginner as unenterprising. If one man's labor can be made to net an income of $1,500 a year, it is a very natural inference that two men could double, or four men quadruple, those figures. But poultry raising has its difficulties and limitations, as is probably the case with every business.

The best eastern poultry journals fix the profitable limit at two thousand birds, and the repeated failures of plants that have started on a much larger scale seem to bear them out. If the editors of these papers were working in 'California they would probably cut down their figures even lower. There are various reasons for this, but an all-sufficient one lies in the difficulty of getting skilled labor, or responsible and intelligent labor, or indeed, labor of any kind. In these days, and especially in this part of the world, where the labor supply is very insufficient, the employer must be prepared to find himself single handed at any hour. In some kinds of work this is not so disastrous to the enterprise, but in poultry raising no detail of the routine can be omitted safely for a day. I have tested the matter thoroughly, and find that with the most careful planning, one thousand laying hens and the younger birds necessary to reinforce them is the utmost limit of one man's capacity.

So it is better not to undertake too much, at least at first, but to consider a plant that will house and yard a thousand layers, and hatch a thousand chickens every year. In the ordinary course of nature about half of these will be pullets, and the other half cockerels. The latter should be sold as broilers when twelve weeks old, and the pullets kept growing until the hens begin to drop off in their laying, which will be toward the end of July, or just before they moult. Arrange to sell off five hundred hens now, clean out their houses, and fill up with the five hundred pullets. These pullets will be laying while the remaining five hundred hens are moulting, and by the beginning of December all the thousand birds should be laying from seven to ten eggs each a month. This average increases up to April, when they should be laying about eighteen eggs a month; after this it will gradually decrease as the moulting season draws near. Of course these dates are only approximate, and sometimes vary as much as six or eight weeks. I am speaking of White Leghorns, as I consider them one of the best breeds for our purpose, and besides, my data concerning them are from my own personal experience, for I am simply telling you what I have done, and am doing now.

Many people talk of the two hundred egg a year hen as though she were an established fact. I find that a hundred and thirty-two eggs a year, is a good average to work on. I have averaged a hundred and forty-six eggs from 900 hens, but prefer to work on the lower number. I would not knowingly hatch eggs from a two hundred egg hen towards the end of the season, for the progeny would naturally be weaklings. The tremendous drain on the vitality of the hen would, in my opinion, finish her in a season as a breeder.

We can reckon on getting eleven dozen eggs per hen per year; these eggs will average twenty-five cents a dozen, or $2.75 a year; allowing $1.20 for feed, this will leave a profit of $1.55 per hen, or $1,550 on a thousand hens. The broilers should sell at forty cents each and cost twenty cents to raise, leaving a profit of twenty cents each, or $100. This gives a total profit of $1,650. The sale of the five hundred old hens will pay for hatching and raising the pullets to laying age. I would consider the $100, from the sale of the broilers as an emergency fund, for all kinds of little expenses are constantly creeping in that cannot be classed as feed expenses.

CHAPTER III.

LAYING OUT THE PLANT

Five acres of ground is about as much as one person can attend to; that is, one acre for the home, lawn, orchard, garden and stable, the other four for the chickens. This land should be selected not too far from the center of a city—if possible, not more than three or four miles, and near a car-line or depot, for convenience in sending in eggs to market while fresh. A daily egg delivery that can be depended on always, is of great importance, for the best prices can only be obtained by keeping the customer steadily supplied every day, with eggs that are absolutely fresh. (Any egg found outside the nest, whose age is in the least doubtful, should go to your own kitchen. This is a golden rule in the poultry business.)

Often arrangements can be made with a good hotel to supply it with eggs at a fixed rate, all through the year. This is an advantageous plan, as a first-class hotel is glad to pay well for eggs that are above reproach; and all the eggs that can be raised on four acres will barely supply an average hotel. The hotel may also take all the superfluous birds as well, but would want them dressed, not alive.

Land can be bought for from $100 to $200 an acre, within a reasonable distance of most of the southern California towns; and keeping in view the increasing value of land, it would probably be wiser to get closer in and pay the higher price.

It is not necessary to get the finest farming land for chickens, but it is as well to get fairly good land, and if some trees are on it, so much the better, for trees are good for shade in the hot summer, and make the home more sightly.

The water question is a most serious one, for without the means of getting water no land is of much value. During the summer five acres planted to home and chickens will require at least 2,500 gallons a day. If there is a private or public water system and means of getting the water piped to the place, it will cost you from fifteen to twenty-five cents a thousand gallons by meter measurement; in some places it may be more. But be sure that 2,500 gallons a day can be forthcoming in the hottest weather, before buying the land.

In many places water can be developed on the land by digging a well from thirty to a hundred feet deep. If others on adjoining land get their water in this way, you may be reasonably sure of being able to do the same, and it is far preferable to buying water from a company. I have tried both ways, and find it cheaper and more satisfactory to be my own water company.

Let us suppose that the five acres of land have been selected within easy distance of the town, at $200 an acre. The land lies sufficiently high to escape floods in the rainy season, and is not so steep that the top soil will wash with a heavy rain. The water supply has been settled, either through a water company, or by having proof that it can be obtained at a reasonable depth by digging a well. If a well is to be dug, it should be done at once, and for information on this subject I refer to the chapter on "Water."

Summer is the best time to buy, as real estate is generally slack at this season; towards the end of the year, when the rains may be expected, it has a tendency to rise. There is also the advantage of having five or six months to get everything settled and in good shape before the hatching season begins.

For convenience, I have made a diagram of five acres in a piece four hundred feet wide and about five hundred and fifty feet long. Of course, the ground may be of quite a different shape, and the yards arranged differently; but if the general idea be grasped, the shape would be immaterial, so long as compactness, accessibility and convenience for feeding and watering are observed.

In laying out your land, the first thing to do is to make a careful and correct outline chart of it. For the home, reserve a space of about 140 x 170 feet, or the size of three city lots, in the southwest corner of the five acres. The prevailing winds come from this direction, and the house being here no unpleasant odors are likely to be wafted towards it. A lawn of about a hundred feet square is planted to white clover, a strip of which can be cut each day and fed to the chickens fresh. There is nothing better than white clover for keeping fowls in good condition and making them lay good, well-flavored eggs. Then, locate barn, store house, incubator house, two brood houses, and twelve small houses and yards for the small chickens. Arrange for all of these to be near the dwelling house, so that a few steps will take you to any of them at any time. The force of this will be appreciated during the hatching season, for the lamps, of which there will be from ten to fourteen constantly burning for months, must be looked at the last thing at night, before turning in, to insure a safe night.

Now, when everything is satisfactorily arranged, go with a tape line and stake off on the ground all that you have marked on the chart. The balance of the land must then be divided into twelve large double yards to accommodate the thousand grown hens. In plotting out the grounds, follow the plans marked in my chart as closely as possible. I have aimed to have all my houses face south, and be flooded with sunlight all day; also, to save footsteps as much as possible, and yet give the birds room enough to thrive in.

Sunlight drives away smells and keeps the houses sweet. When birds are kept under these conditions they are easily kept free from lice by spraying the perches once a week with a strong disinfectant, and by keeping a small piece of ground moist and soft for a dust bath. I claim that by my system of housing and caring for the birds, the old fashioned plan of dusting each bird with insect powder is absolutely unnecessary; and to prove it I can truly say that my hens are free from lice and mites nearly, if not all the time. I have picked up hens constantly, and after a most careful search have found nothing. This state of freedom from insects is possible only by using incubators and brooders, for a setting hen will breed and impart them to her chicks, no matter how often she is dusted. In a warm climate like ours one of the chief causes of weak hens, and, consequently, poor laying qualities, is insects.

We will now suppose that the ground is laid out and ready for fencing and building. Before going any farther, a few remarks on the materials to be used and the labor to be employed will be in place.

After five year's use I find that all my houses with cement floors, fences, posts, etc., are in perfect order and no expenses for patching or renewing have been incurred in that time, except for painting and whitewashing. But where I have used wood floors, I have just had to replace them with cement. This convinces me that it is better to do good work in the beginning, and so have something that will last without tinkering. for fifteen or twenty years or even more. My wooden floors would have lasted a good deal longer if I had built them higher above the ground; but that plan has serious objection, for it leaves a space for the accumulation of dirt and rubbish; and the hens are fond of laying in such places, which necessitates one's crawling in after the eggs, and a painful loss of time and temper. Sometimes, during the rainy season, water gets under these floors and makes things highly dangerous for the fowl's health, besides creating bad odors. I am therefore a believer in cement floors, lightly covered with sand or dry soil; the droppings are then easily removed each day and a handful or so of sand thrown over the floor again. In the brooders, with so many lamps burning, cement is safer than wood. Again, in store room and feed house, cement is a safeguard against rats getting in through the floor.

I use surfaced wood for my buildings. I find it pays to have the rafters, studding and plates of surfaced Oregon pine, and the sides of rustic which is surfaced on one side only. The doors are of flooring, which makes a stronger job than redwood. Surfaced woods are more easily handled, the joints fit better, and the saw stays sharp a good deal longer. It irritates one to see a carpenter, at $3 or $4, spend ten minutes several times a day sharpening saws, which is usually the case when rough lumber is being used.

Sawn shakes, 12" to weather, are just as good as shingles for roofing, if they are painted as soon as laid. Here again we can save a good deal of time, as the average man can cover twice as fast with shakes than with shingles.

I paint everything outside just as soon as the carpenter finishes a house, and whitewash the inside and cement floor as well. This painting, before the wood dries out too much, keeps it in good shape indefinitely. I have shake roofs treated in this manner, that after five years do not show a split or a curled edge. I use good paint for the first coat, and afterwards paint every two or three years.

Some may consider my methods rather expensive for chicken houses. But experience proves that a little more in the initial cost is money saved later on; and besides, a neat, well built house is much easier to keep clean. Here in Southern California it behooves us never to lose sight of the fact that labor is very dear, very scarce, and not very good at the best. A new country affords so many openings for the average man that the employer often has to take whatever offers, and be thankful for a poor bargain. Sometimes one finds it impossible to get any help, and has to turn to and do everything alone, and at such times he appreciates the blessing of a convenient and well regulated plant. I think my plant pretty good, and were I building again I would have it still better; for nothing is more disheartening than constant renewals and repairs. No, it is safer to finish with capital at the beginning and consider all profit as profit, not to be broken into for what should have been done thoroughly at the start.

For the first six months, while the plant is building, it pays handsomely to employ a good, all-round man; one who is handy with tools, who uses his brain as well as his hands, and especially, one who will take an interest in his work. Pay him well and treat him with consideration, and he will soon get the run of the ropes, and you will not have to be continually changing, and with each change having to explain everything over again to the new man.

CHAPTER IV.

BUILDINGS AND SPECIFICATIONS

I would have a carpenter build the two brood houses, the store house, one hen house and one chicken house. The handy man can build the rest of the hen houses and chicken houses from the carpenter's models. You would then be sure of having all your houses built on correct principles.

The first thing to build will be the store house, which can be used all through the building operations for storing the hardware, tools, paints, etc. From the beginning, insist on having everything portable under lock and key each night; no time is then wasted in the morning looking for things, which means quite a saving of time in the course of a year.

A corner of the house can be partitioned off to make a room in case a man is hired who will live on the place; or the family can live in it with the addition of a few screens, while the residence is being built. Very little grain will be required for the first nine months after the plant is started, so the house will be comparatively empty.

To the novice, this house will appear rather, and unnecessarily, a big expense; but as time goes on, the comfort and necessity of it will be more and more felt. The money saved in one year, by buying the year's supply at the right time instead of in small lots, will more than pay the whole cost of the house. This is one of the economies that greatly adds to the success of the chicken business. A good start with a well built and systematically arranged plant means less time lost in tinkering, and therefore more time available for caring for the birds.

The store house must be rat-proof, so a cement floor will be best. It will be one story, and 20x30 feet, large enough to store the grain necessary for feeding a thousand birds for one year, and the usual odds and ends belonging to the plant. Lay it off as near the road as possible, for the convenience of hauling stores; then, after watering the ground at least eighteen inches deep and firming it by tamping, get it fairly level; but be careful not to lower it below the general contour of the land, for we do not want a basin for the water to lodge in during the rainy season. When level, take some 3"x4" sawn redwood posts ten feet long, and lay them in the form of a frame around the outside edge of the foundation, with the 4″ side of the wood touching the ground. After leveling them with a spirit level, drive wooden pegs in every two feet, on the outside of the posts (Fig. 1). We now have a frame ready for the cement floor,

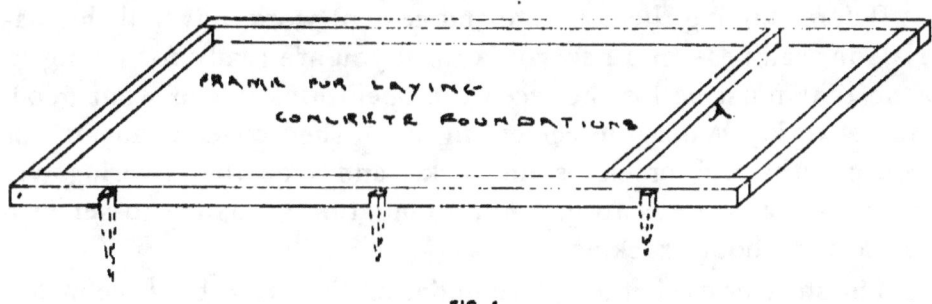

FIG. 1

which, when filled with concrete and smoothed over with cement, will give us a floor three inches thick. Having the frame well pegged and perfectly level makes the tamping and leveling of the concrete easy work. It will be well to remember that a little time spent on having all the frames for foundations perfectly level, and the corners square, will be time well spent, for uneven and crooked foundations make slow and unsatisfactory work when building the house.

Make a rough box about four feet wide and six feet long, and six inches deep, for mixing up the concrete and cement, and do all the mixing alongside the foundation.

The usual way of making concrete is one part of good cement to six or seven parts of broken stone, gravel, or slag. Mix dry, and when ready, spray water on gently, at the same time turning the materials over and over, until the mass will hold no more water, and yet retain the water already in it. Now shovel it into the frame until it is level with the top, then block it up with the piece of post A (Fig. 1) and commence tamping it with a regular hand tamp. Tamp it all over until it is hard and level, and about half an inch below the level of the frame. Now move the post A back another foot, and stake it in firmly, ready for the next batch of concrete. After the first batch you will know exactly how far back to put the wood.

When the concrete is in, mix one part of cement to two parts sand (be sure to get sharp sand, not sand with the edges rounded off), and enough water to stay in the mixture and not run out. Be careful to mix the cement and sand thoroughly before adding the water. Lay it on the concrete, the same as before, level it with the top of the frame, and then smooth it off with a steel plastering trowel. With the constant smoothing a little water will settle on top of the cement; shake a little dry cement on, and work it over until it gets a smooth, glossy surface.

Before starting the concrete and cement work, have all the materials in readiness, and never mix until you are ready for laying it, for no time must be lost between the operations if you want good, durable work. When the cementing is finished cover it an inch or so deep with sand, or else some sacks, and keep the covering wet for several days, so as to give the foundation a chance to set firm and hard, without cracking.

The store house is the only foundation that need be three inches deep, for it has to stand the strain of a good many tons of grain. The other houses can have the frames filled and tamped one inch deep with soil, before putting in the concrete and cement. This would give about one and three-quarters inches of concrete and a quarter inch of cement. If the work is done properly and quickly, a quarter inch of cement will hold just as firm as an inch; all that is required is a smooth, glossy surface that will not hold dirt.

The best plan for one who is not used to this sort of work would be to get the frames of the foundations for the store house and two brood houses laid and levelled, then get a regular cement sidewalk layer to come out and contract for concreting and cementing them. He will charge from seven to ten cents a square foot for laying them, according to the distance he has to haul materials, and the thickness of the cement you want laid on the concrete. You can then watch and note how he does it, and the proportions he uses, where he gets his stone or gravel from and how much it costs per load. He will, if decently treated, tell you how to lay the rest of the foundations yourself, and in the cheapest way. A dollar or two thrown in over the contract price, and a pleasant manner, will mean a lot of information on the subject, that will save many dollars in the construction of the rest of the foundations and make a much better job of them.

It is better to put a little more cement than concrete around the edge of the foundation, close to the frame, so that when the frame is removed cement, not concrete, will be exposed to the weather. (Fig. 2)

The foundations and frames should be kept wet and not disturbed for at least ten days, after which the frames can be removed. The redwood posts used for the frames will afterwards be used for fence posts.

When finished the foundations should be at least three inches above the level of the surrounding ground, or else a lot of trouble will be experienced in having to make drainage ditches around the houses in wet weather. Dry houses, all the time, are absolutely necessary for the health of the birds.

In building the store house (Fig. 3) lay the plates on top of the cement, and when the frame is up let the rustic overlap an inch or two over the cement. This will make it rat and water proof. (Fig. 2)

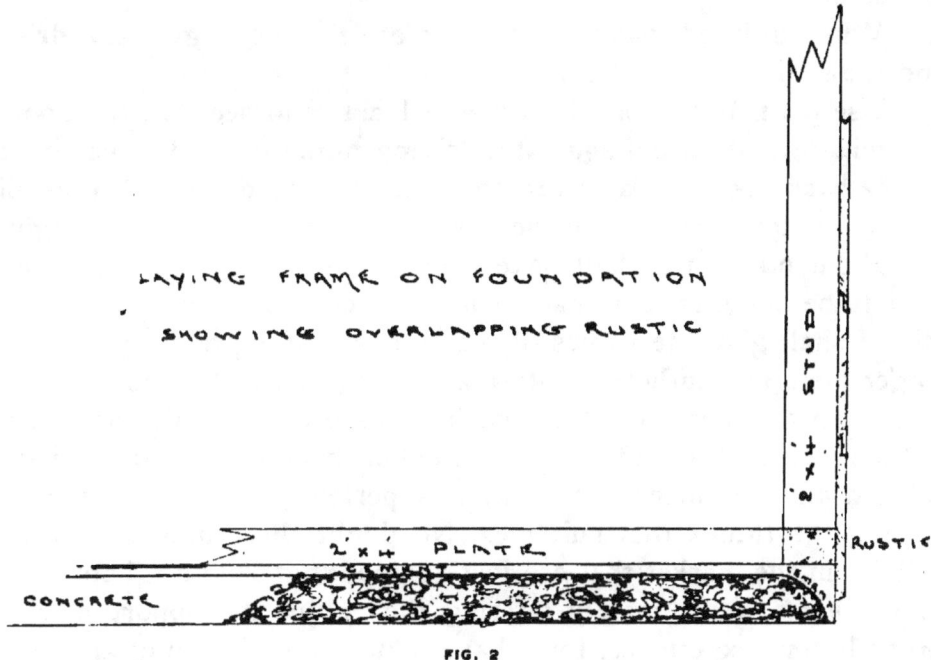

FIG. 2

Use surfaced wood on the four sides; 2"x4" Oregon pine for plates, studs and rafters, and 8" rustic for the outside; sawn shakes 12" to weather for roof; sheeting 1"x4" Oregon pine; studding and rafters about three feet from center to center. All of the houses are made from these materials and measurements. Doors are cheaper bought ready made than if made by yourself or a carpenter. Where windows are required, instead of putting the sashes one over the other, put them side by side, on hinges opening in, for all the houses are low.

The store house will need one full sized door in front and one behind, and about four sets of two sashes side by side, a bench and a few shelves close to the back door; but all the rest of the house should be bare, for the stowage of the grains, etc. I keep all my grains in separate piles, receiving them through the front door and using them from the back door.

Arrange windows and front and back doors the same, so that the space between the two doors is kept free, leaving the two sides for stowing the sacks. No windows are required at the ends of the house, for the sacks will have to be stowed high up against the walls. Always keep "cat space" between the walls and the sacks, in case of a damp wall, or water getting in. Fill up every little space between the sheeting and anywhere else that might let a mouse in.

When finished, paint at once, before the wood gets sun dried and the shakes curl and split.

Use plain but strong hinges and locks, and see that the doors and windows are proof against a driving rain. A good screen door on the back door, and a screen on one of the windows, will give all the ventilation necessary in the day time. None is required at night.

From now on I shall have occasion to mention articles that have to be bought ready made, such as incubators, fence wire, etc., etc. I shall give the names of the things that I prefer and where to get them in Southern California. It is not usual to do this, unless as an advertisement, so I wish to state that in this case I am not paid for anything that is mentioned in this book. In fact, I have refused advertisements, for I want a perfectly free hand in mentioning the things that suit me. No doubt there may be similar articles in the market that are just as good, but I do not happen to know them. I am merely offering the novice the opportunity to profit by my experience, for I have done a lot of buying and condemning, until I got what I considered the best. My plant being complete, there is not much likelihood of my buying anything more, and asking for a reduction on the strength of this book.

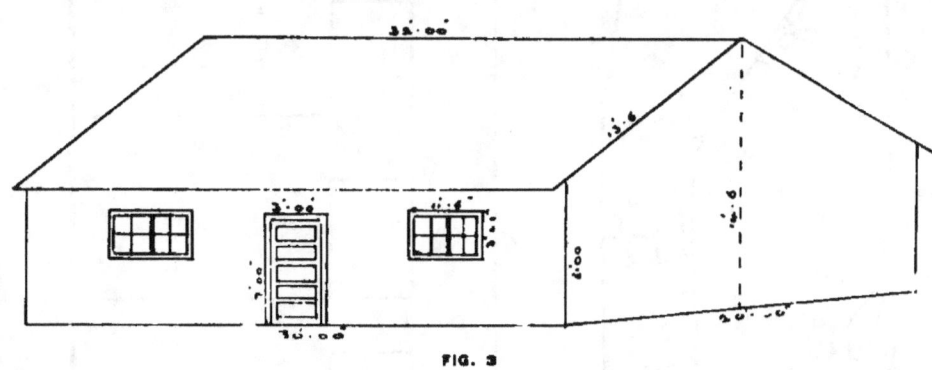

FIG. 3

Specifications for Store House. (Fig. 3)

Cement floor, 600 square feet @ 7c per foot..............$ 42.00
Plates and Braces,*15—2"x4"x20' s4s O. P. 200 ft.
Rafters,   22—2"x4"x14' s4s O. P. 200 ft.
Studs,   25—2"x4"x16' s4s O. P. 420 ft.
Sheeting,   .60—1"x4"x16' s4s O. P. 330 ft.

_____

                              1150 ft. @ $24..... 27.50
Surfacing above 6.00
Rustic, 90—1"x8"x20' No. 2 @ $30...................... 36.00
Window and Door Sill, 22 ft. of 2"x8" 1.50
Casements, etc., 7—1"x5"x20' s4s Redwood @ $40........ 2.50
Sawn Shakes, 75 bundles @ 40c per bundle.............. 30.00
2 Colonial Doors, 1½"x3'x6' 6" @ $3.50 each........... 7.00
4 Windows, 27"x44", 12 lights, @ $1.50 aech.............. 6.00
Paint, Hardware, etc.................................. 36.50
Carpenter, 10 days @ $3.50 per day..................... 35.00

_____

                                       $230.00

*Overhead joists to keep the walls from bulging out.

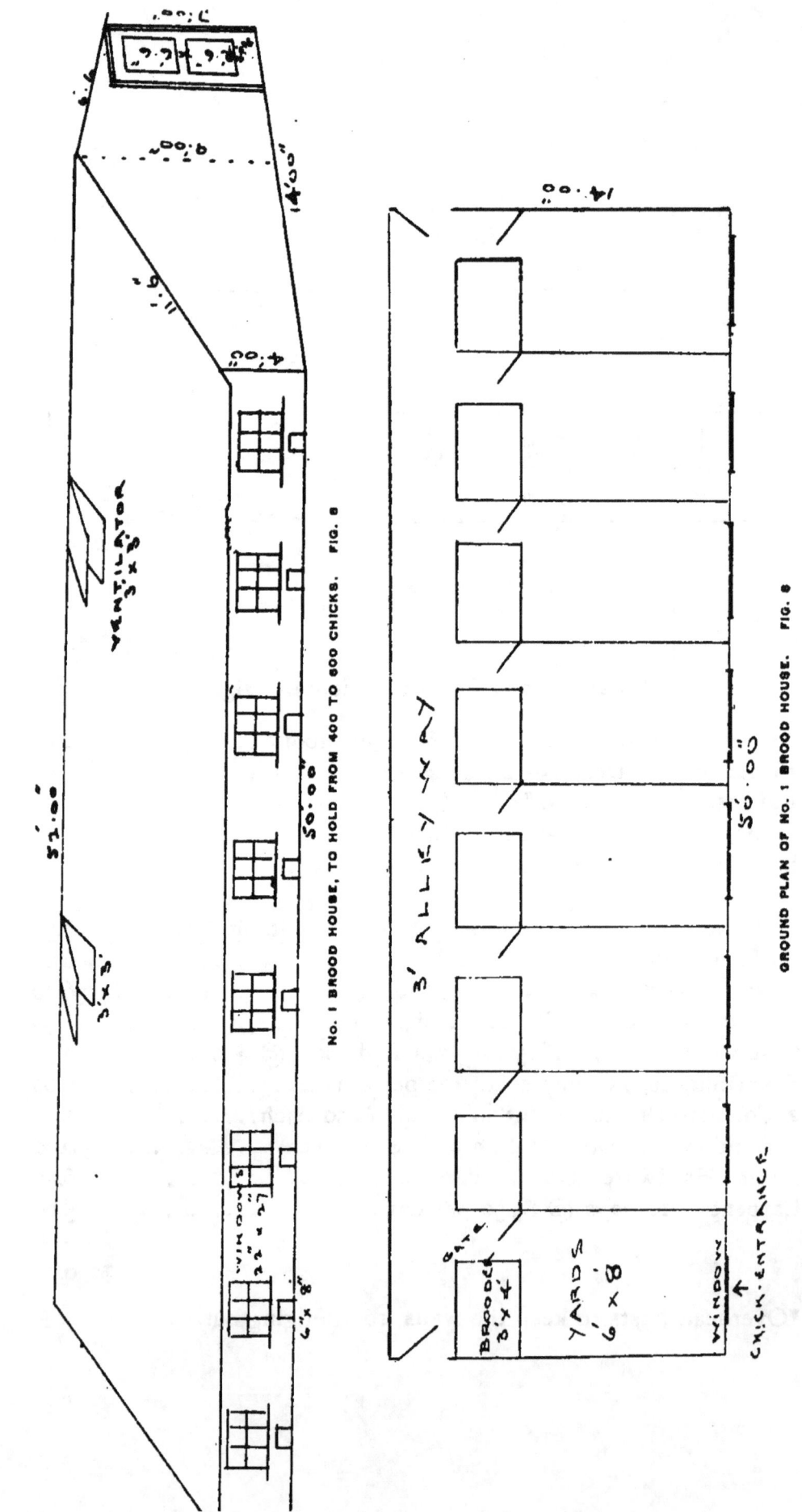

VENTILATOR
3 x 3

3 x 3

VENTILATOR

NO. 1 BROOD HOUSE, TO HOLD FROM 400 TO 600 CHICKS.   FIG. 8

3' ALLEY WAY

BROODER
3 x 4

GATE

YARDS
6 x 8

WINDOW

CHICK ENTRANCE

GROUND PLAN OF NO. 1 BROOD HOUSE.   FIG. 8

Specifications for Brood House No. 1. (Fig. 8)

Cement Floor, 700 square feet @ 7c per foot...............$ 49.00
Plates,　　　　　12—2"x4"x20' s4s O. P. 160 ft.
Studs,　　　　　35—2"x4"x12' s4s O. P. 280 ft.
Rafters,　　　　18—2"x4"x18' s4s O. P. 220 ft.
Sheeting,　　　　50—2"x4"x22' s4s O. P. 350 ft.

　　　　　　　　　　　　　　　　1010 ft. @ $24..... 24.00
Surfacing above 5.00
Rustic, 60—1"x8"x20' No. 2, 815 ft. @ $30.............. 24.50
Casements, 6—1"x5"x22' s4s Redwood @ $40............ 2.40
2 Colonial Doors, 1¼"x2'6"x6'6" @ $2.40 each 4.80
4 Windows, 27"x44", 12 lights, @ $1.25 each............. 6.00
Sawn Shakes, 80 bundles @ 40c........................32.00
Paint, Hardware, etc................................... 17.30
Carpenter, 10 days @ $3.50 per day..................... 35.00

　　　　　　　　　　　　　　　　　　　　$200.00

Window sills can be made from the 1"x5" redwood for casements.

Nail a piece of coarse screen wire inside of ventilator opening to keep out cats, etc.

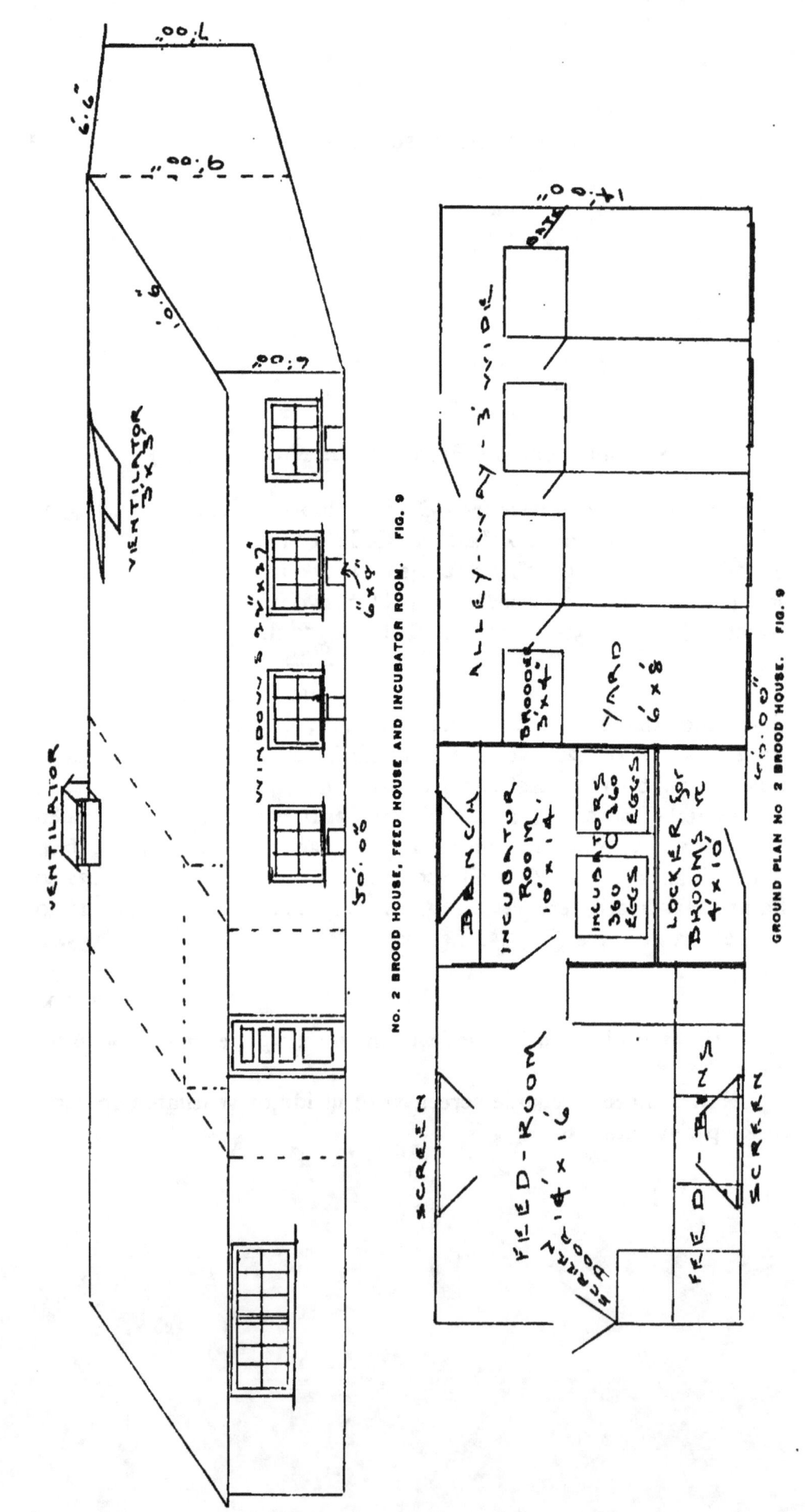

VENTILATOR

VENTILATOR 3'x3'

WINDOWS 24"x27"

6'x6"

7'.00"

10'.0"

9'.0"

9'.0"

9'.6"

7'.00"

NO. 2 BROOD HOUSE, FEED HOUSE AND INCUBATOR ROOM. FIG. 9

ALLEY-WAY - 3' WIDE

BROODER 3'x4'

YARD 6'x8'

BENCH

INCUBATOR ROOM. 10'x14'

INCUBATORS 360 EGGS

LOCKER for BROODS &c 4'x10'

SCREEN

FEED-ROOM 14'x16'

DOOR

FEED-PANS

SCREEN

14'.00"

4'.0 0"

GROUND PLAN No 2 BROOD HOUSE. FIG. 9

Specifications for No. 2 Brood House With Incubator Room and Feed Room. (Fig. 9)

Cement Floor, 700 square feet @ 7c per foot..............$ 49.00
Plates,   15—2"x4"x20' s4s O. P. 200 ft.
Studs,   40—2"x4"x14' s4s O. P. 400 ft.
Rafters,   18—2"x4"x18' s4s O. P. 220 ft.
Sheeting,   50—1"x4"x22' s4s O. P. 350 ft.

                                    1170 @ $24........   28.80
Surfacing above 6.00
Window and Door Sills, 24 ft. 2"x8"..................   1.50
Shelves, etc., 5—1"x12"x20' s4s Redwood..............   3.00
Rustic, 80—1"x8"x20' No. 2, 1080 ft. @ $30......... 32.40
Casements, etc., 6—1"x5"x22' s4s Redwood............   2.40
*Ceiling, 24—1"x6"x14' No. 2 Flooring................   5.00
Colonial Doors, 4—1¼"x2' 6"x6' 6" @ $2.40 each........   9.00
Windows, 5—27"x44", 12 lights, @ $1.50 each...........   7.50
Sawn Shakes, 80 bundles @ 48c each.................... 32.00
Paint, Hardware, etc.................................. 27.00
Carpenter, 10 days @ $3.50 per day.................... 35.00

                                                      $240.00

*Incubator room to be ceiled **overhead** only.

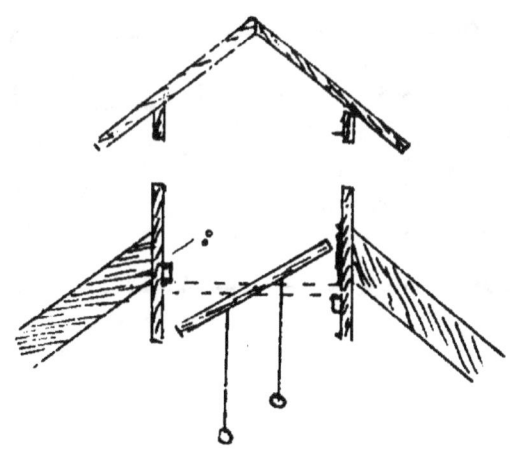

DIAGRAM OF INCUBATOR ROOM VENTILATOR

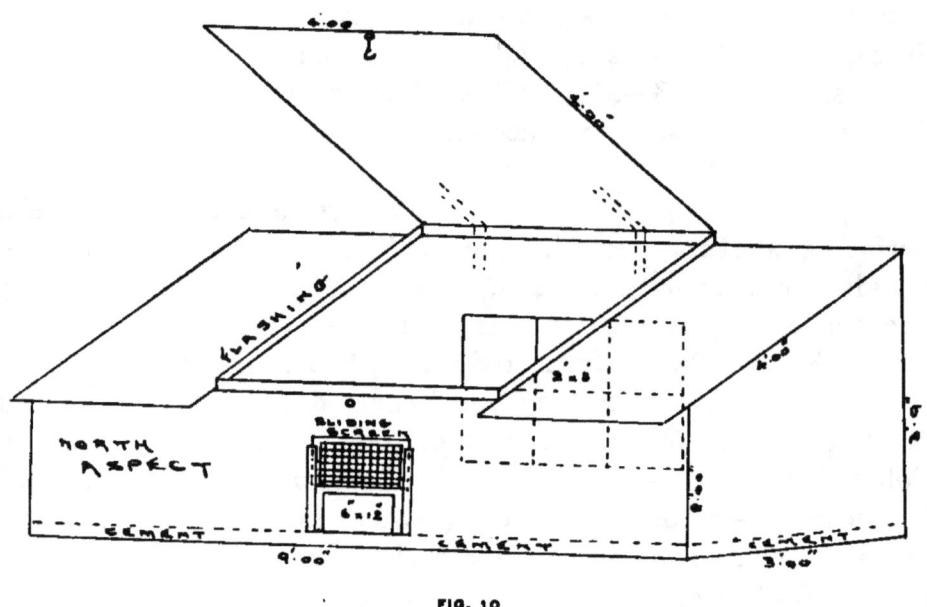

FIG. 10

FIG. 11

Specifications for Chicken House. (Fig. 10)

```
Cement Floor, 27 square feet @ 7c per foot..............$ 1.90
Frame, 10—1"x4"x10' s4s O. P.
Perches, 3—1"x2"x12' s4s O. P.—50 ft. @ $26.............  1.30
Rustic, 12—1"x8"x12' No. 2 @ 30.......................  3.00
1 Sash, 22"x27", 6 lights..............................   .75
Sawn Shakes, 4 bundles @ 40c per bundle................  1.60
Flashing .............................................   .25
Paint and Hardware (2 strong strap hinges)............  1.20
                                                        _____
    Complete .........................................$10.00
```

The flashing is important and goes under the shakes around the frame of flap to prevent leaks in rainy weather.

The window is a fixture, as all the ventilation necessary comes through the roof, the flap of which is propped up with a perch at cleaning time, and left open all day. When closing the flap at night fasten it down with a hook, as a precaution against its being blown open at night by a gust of wind.

After the house is built lay it on its back and thoroughly tar around the bottom where the cement will go, and when dry, place it on its foundation, which has been previously watered and tamped level and hard, drive in some stakes around the outside, to keep the house in shape while tamping the floor, the concrete and cement can then be laid. I find this plan of laying the cement floor after the house is made and using the house itself for a frame, very convenient with small houses which can be easily moved.

The small door in front will give all the ventilation necessary at night, in fact on a cold rainy night I use a wooden slide over this opening unless the chickens are over nine or ten weeks old.

Specifications for 12 Chicken Yards. (Fig. 11)

```
Sawn Posts, 224—3"x4"x8' Redwood @ 20c each .........$ 44.80
Gates (12), 24—1"x4"x12' s4s O. P., 100 ft. @ $26..........  2.60
100 rds 6 ft."Union Lock" Poultry Fence Wire @ 60c per rd. 60.00
Paint, Distillate, and Coal Tar .........................  4.00
Hinges, Spikes, Staples and Hooks ......................  2.00
12 Water Tins, 2 gallons each and shallow .............  6.00
12 Grit Jars @ 10c each ...............................  1.20
36 Feed Troughs, 4 feet long and shallow ...............  12.00
                                                        _____
                                                       $133.00
```

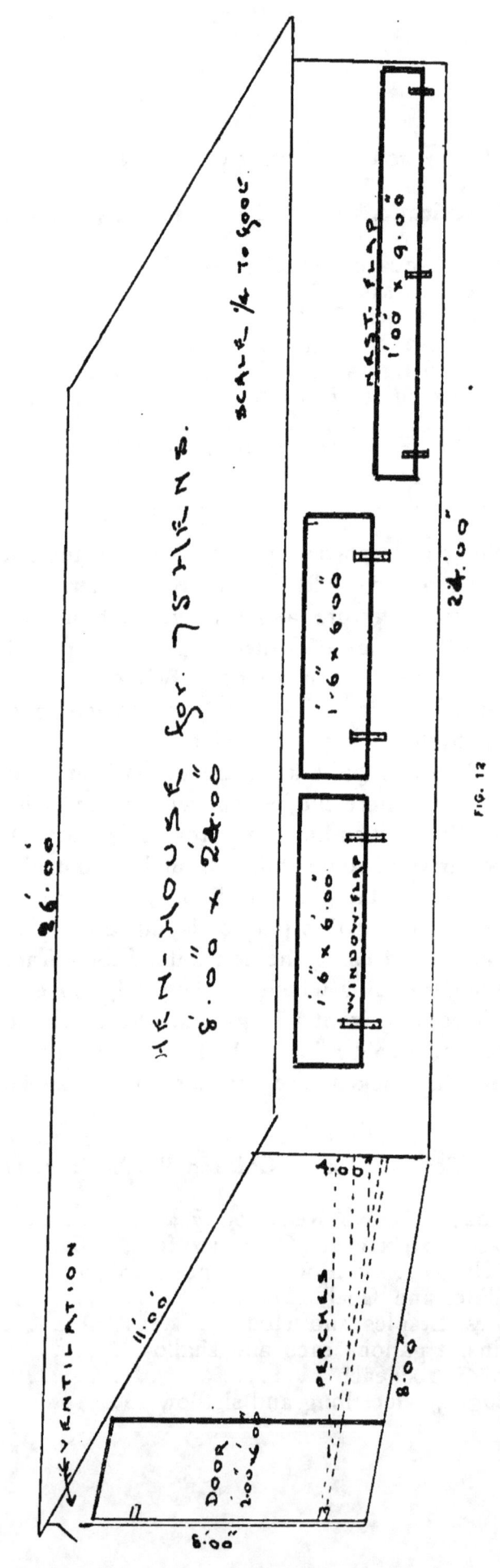

HEN-HOUSE for 75 HENS.
8.00″ x 24.00″

SCALE ⅛″ TO 6.00′

NEST-FLAP
1.00′ x 9.00″

1.6″ x 6.00″

1.6″ x 6.00″
WINDOW-FLAP

24.00″

36.00′

VENTILATION

11.00′

PERCHES

4.00′

8.00′

DOOR
2.00 x 4.00

6.00″

FIG. 12

Specifications for Hen House. (Fig. 12)

Cement Floor, 192 square feet @ 7c per foot..............$13.45
Plates and Perches, 9—2"x4"x24' s4s O.P. 144 ft.
Studs,   15—2"x4"x12' s4s O. P. 120 ft.
Rafters,   9—2"x4"x12' s4s O. P. 72 ft.
Sheeting,   13—1"x4"x26' s4s O. P. 120 ft.—456 ft. 11.00
Surfacing above 2.00
Rustic, 1"x8"x24' No. 2, 480 feet @ $30................... 14.00
*Doors and Flaps, 12—1"x6"x12" No. 1 Flooring, 72 ft. @ $40   2.90
Sawn Shakes, 24 bundles @ 40c per bundle................ 9.60
Paint, Hardware, etc................................... 8.05
Nests 2.30
Jail 1.30
                                                        ———
                                                      $65.00

*Doors and flaps are best made from pine flooring, which is much stronger and holds together better than redwood.

Nests.

Bottom boards, 1—1"x12"x10' s4s Redwood.
Bottom boards, 1—1"x3"x10' s4s Redwood.
Side boards, 1—1"x12"x14' s4s Redwood.
Side boards, 1—1"x6"x10' s4s Redwood...................$1.10
Lid boards, 6—½"x6"x10' pine ceiling.................... 1.05
Hinges and nails .... .15
                                                        ———
                                                      $2.30

Jail.

Frame, 1—2"x4"x16' s4s O. P.
Frame, 4—1"x2"x10' s4s O. P. $0.50
Grating, 18—½"x2"x6' surfaced battens .... .30
1 Bundle Sawn Shakes .... .40
1 Pair Small Strap Hinges .... .10
                                                        ———
                                                      $1.30

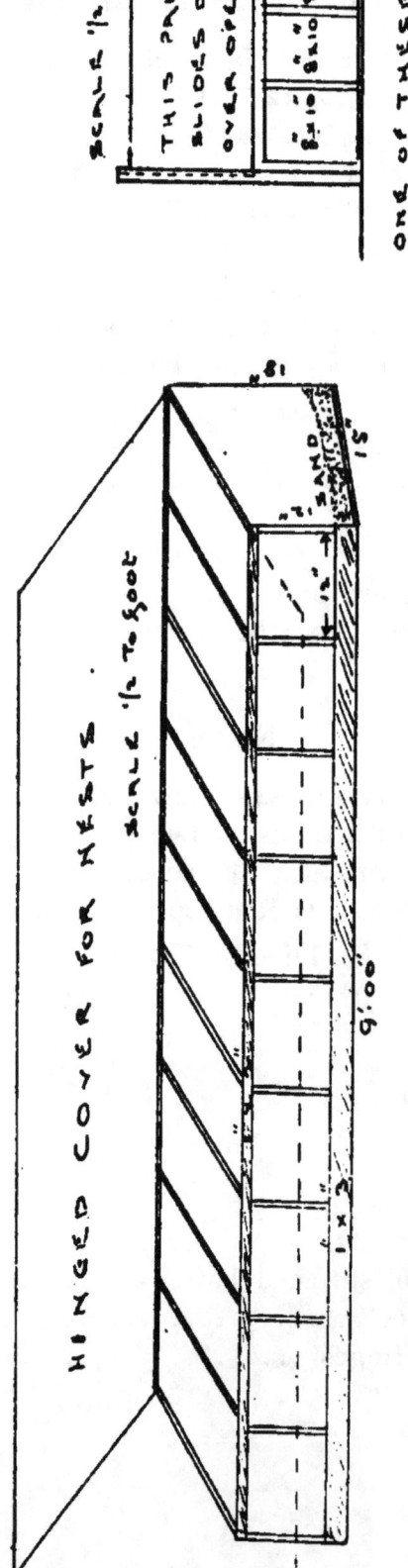

SCALE ½ TO FOOT

THIS PANEL
SLIDES DOWN
OVER OPENINGS

8x10 8x10 8x10

ONE OF THESE SETS OF
OPENINGS LEADS INTO EACH
YARD FROM BACK OF
HEN-HOUSE

HINGED COVER FOR NESTS

SCALE ½ TO FOOT

9'·00

1 x 2

SAND

NEST- BOXES

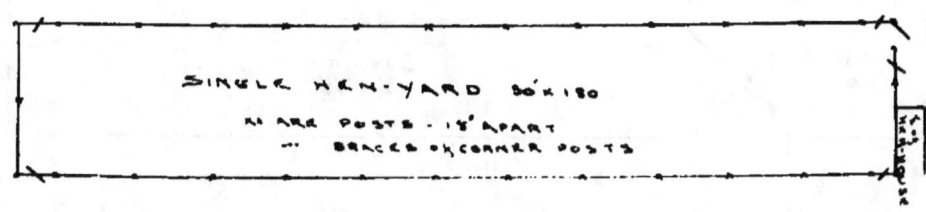

FIG. 13

Specifications for 12 Double Hen Yards. (Fig. 13)

Each single yard being 30x180 feet.

Sawn Posts, 54—6"x6"x8' Redwood @ 60c each ....$ 32.40
Sawn Posts, 500—3"x4"x8' Redwood @ 20c each......... 100.00
Gates (24), 48—1"x4"x12' s4s O. P.—200 ft. @ $26......... 5.20
400 rds. 6 ft."Union Lock" Poultry Fence Wire @ 60c per rd 240.00
1000 ft. Soft No. 9 Wire for bracing corner posts........ 3.40
Paint, Distillate and Coal Tar.......................... 12.00
Hinges, Spikes, Staples and Hooks..................... 4.00
*Water Tins, Feed Troughs and Grit Boxes............. 33.00
                                                    ————
                                                    $430.00

*Each of the twelve yards will require
    1 3-gallon Galvanized Iron water Tin..$0.75
    4 Feed Troughs, each 6 feet long...... 1.25
    1 Grit, Shell and Charcoal Holder..... .75
                       ————
               $2.75 or $33.00

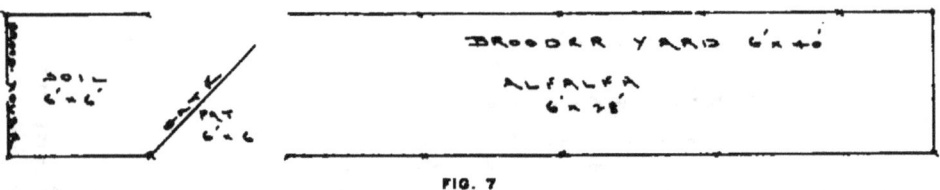

FIG. 7

*12 Brood House Yards. (Fig. 7)

The photograph will give an idea of how the yards in front of the brood houses are arranged. Mine are rather small so I give the dimensions for larger ones.

As they are close to the house the posts are surfaced and all the work finished as neatly as possible. No bracing posts are needed, as the base board and top rail make them unnecessary.

The gates are fastened with ordinary hooks and screw eyes, and are made of 1"x4" pine, with a bracing piece running diagonally from corner to corner. Let the nails go through the wood a quarter of an inch and then clinch them, otherwise the constant jar of opening and shutting the gates, will in time loosen and draw the nails. (Fig. 6)

Posts, 112—3"x4"x6' s4s Redwood @ 20c each $22.40
Base Boards and Gates, 56—1"x6"x12' s4s O. P............. 10.00
Top Rails and Gates, 78—1"x4"x12' s4s O. P.............. 8.40
40 rods "Union Lock" Poultry Wire @ 50c per rod......... 20.00
Paint, Hardware, etc. 9.20

$70.00

* The general chart gives only nine yards, which is a mistake.

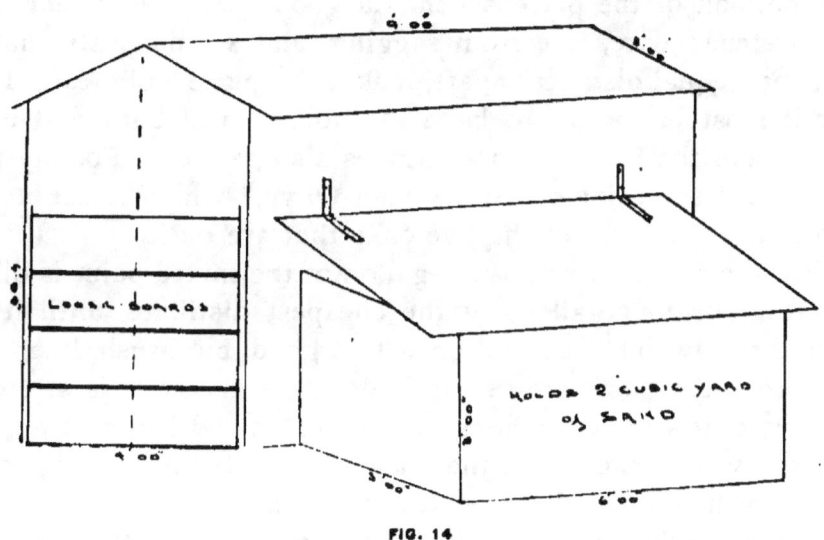

FIG. 14

Compost House and Sand Box. (Fig. 14)

The lumber for building the two compost houses need not be bought especially for this purpose, as there will probably be enough pieces left over when all the building is finished to build these small houses, a coat of paint will cover any patchwork that may be necessary in order to use up the small pieces.

The two sand boxes should be made from 2x12 redwood, as they should be made strong. They will take 6 pieces of 2″x12″x12′ rough redwood and cost about $4.50.

CHAPTER V.

FENCING

The look of the place depends a good deal on the fences. A well stretched wire, free from sagging, and set on posts that are plumb and equal distances apart, makes the place look neat. I consider it most important to have the houses well built and neatly painted, and the fence posts and gates also painted. For my place I have used the color called "Golden Olive," which is the nearest approach to the color of the live oaks that are on it.

For the rough sawn posts a gallon of the mixed paint is diluted with three or four gallons of the cheapest distillate, until it just colors the redwood. It is put on with a good, big brush, like whitewash, after the posts are set up and before the wire is stretched. This one coat will last as long as the posts; and if the tops of the posts are well soaked with the wash, few of them will require renewing in the course of ten or fifteen years.

A few weeks before the fencing begins, get all the posts that will be needed and thoroughly tar one end for a distance of two feet up. When dry, which will be in about two weeks, give them another coat, filling in all cracks, and especially the sawed off part; otherwise, in wet weather the water will creep right up the heart of the post through capillary attraction, and in time rot it away. I like to have a small fire burning under the tar pot, and put on the tar boiling; it penetrates the soft, porous redwood better than if applied cold. When the second coat is dry the posts are ready for setting up.

All corner posts should have an anchor spiked on to keep them firm under the heavy strain of the stretched wire. They should then be braced in two directions, as shown in the diagram. (Fig. 15)

A wire stretcher made especially by the wire manufacturers should be used for stretching the wire. Several hundred feet can be stretched at one time, and only two men are required. (Fig. 16)

I have tried several kinds of wire, but have found nothing to equal the Union Lock wire, made by the Union Fence Company, DeKalb, Illinois. It is simply impossible to make a first class job of wire that has a diagonal mesh, on ground that is not perfectly level. The Union Lock has all the wires either horizontal or perpendicular, and the horizontal wires are closer at the bottom to keep small chicks from getting through. I have some fences over three hundred feet long that are as tight as harp strings the whole distance. Short lengths, up to twenty or thirty feet can be stretched hand taut. A tight wire makes a better support for the posts, and keeps them from wobbling or getting loose in wet weather.

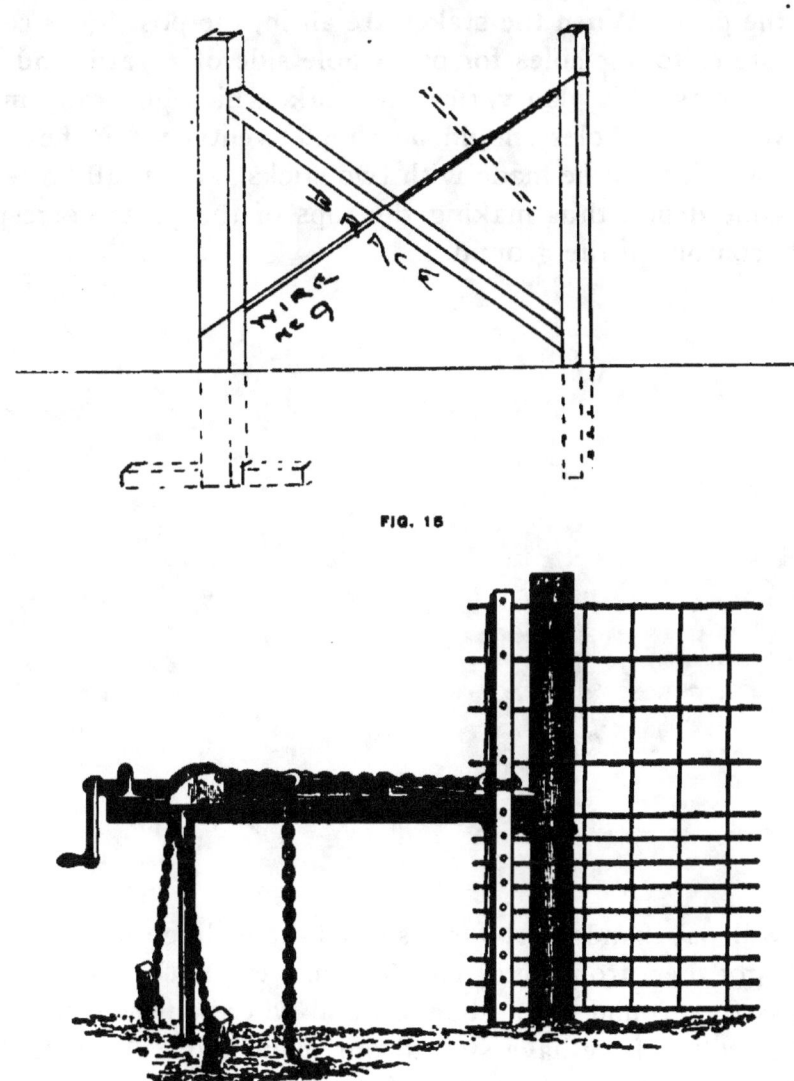

FIG. 15

FIG. 16

The cost of this wire is slightly higher than ordinary poultry wire with octagonal or sextagonal mesh, but it requires no bottom or top rail. This last item costs more than the difference in the price of the wire, and is absolutely necessary where common poultry fencing is used.

4

When ready to begin fencing, which should be the last thing
done, lay off the corners of the land with stakes. Then stretch a
line tight from corner to corner, and drive in a stake every fifteen
feet along this line, except the first stake from each corner post,
which should only be about six feet; this is called the bracing post,
and holds one end of the brace, the other end being on the upper
part of the post. When the stakes are all in, the post holes can be
dug. I prefer to dig holes for one whole side of a yard, and then
set in the posts; this plan varies the work, which generally means
better work. The holes should be about twenty-two inches deep,
not less; a guide can be made with two sticks, so that all holes will
be the same depth, thus making the tops of the posts correspond
with the contour of the ground.

It will pay handsomely to use post-hole diggers. I use two
kinds (I got them from Peter Henderson, New York) and find both
useful, as the ground often changes, and where one will not work,
the other will. If we have started our plant about August, it will
be close to the end of the year before the fencing begins, and the
first rains will probably have fallen, making the ground soft and in
good shape for the digging. Do not try to fence until the ground
is in this condition, for moist soil is necessary to good firm tamping.

When the holes are dug, set the corner or end posts in first;
have them perfectly plumb, and fill up the holes gradually, tamping
hard all the time. If the hole is filled at once and then tamped, the

top will be hard and the bottom soft; so tamp from the bottom up.
A good tamp can be made from a six foot piece of inch and a quarter
iron pipe with a cap screwed on the end. The corner or end posts,
in particular, should be made very firm, for on them comes all the
strain while stretching. While tamping, use the spirit level and
keep your post in line, and straight with the corner posts.

After the brace is spiked on to the corner and bracing posts,
pass a piece of soft wire, size No. 9, around the posts and staple it
to the corner and bracing posts; then put the handle of the ham-
mer through the wire above the brace, and twist the two parts tight.
Do the same below the brace. The turns will stay in and keep
every thing taut. (Fig. 15)

Union Lock wire can be stretched on the frames of the gates
before they are hung; but be sure to have all the painting finished
before putting on the wire. Trying to paint wood covered with
wire is dirty and slow work. The gate posts should all have headers
to keep them from pulling apart through the strain of the wire.
(Fig. 6)

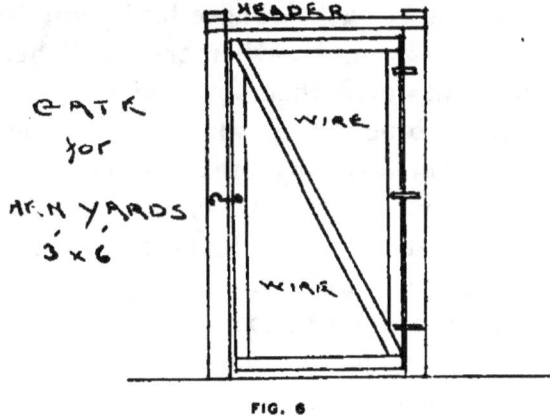

FIG. 6

CHAPTER VI.

HATCHING

There are two ways of getting eggs to start the incubators the first season.

The first is to give all your time and attention to building the plant, and then in February to buy the eggs for hatching, by the incubator lot. These eggs will cost about $10 per hundred if bought from a first class reliable man, whose hens are mature, and great care should be taken to get eggs from thoroughbred stock, that have been bred for size and laying qualities in preference to show and fancy points. A great deal of inbreeding is done to propagate birds for show purposes; but this inbreeding has a tendency to weaken the stamina and laying qualities of the birds.

To hatch a thousand chicks the first year, about twenty-five hundred eggs will be required. If all these are contracted for from one breeder a considerable reduction in the price per hundred can be looked for.

The other way would be to build, say three of the hen houses the first thing, and in August buy about a hundred and fifty hens that have just finished their second season, and are beginning to moult. At this time good birds can be arranged for from a first class breeder at about $1 each. These birds will do little else than eat and moult until November, when they will begin to lay fairly well. On no account force them by giving them stimulants, for you want their eggs to be strong and natural when the incubating season comes. I never force my breeders, as one good, fertile egg is worth three weak ones at that time. The care and feed of these birds, added to the cost of them will come to nearly as much as the cost of buying eggs for hatching. But a good deal of experience can be gained through them in four or five months, so that when the young ones come you will not be quite unversed in chicken lore.

The conditions of the work must decide the course to follow. If you are going to try to do all, or nearly all of the building alone, then by all means have no birds around until the plant is complete; but if you have a good man, and can afford to have a carpenter do some of the building, the second plan is the best.

The incubator room faces north, and has a ceiling; it has a locker to the south of it for utensils and odds and ends, and is wedged between the No. 2 brood house and the feed room; all of these act as double walls and keep the temperature even, which is very necessary to successful hatching.

I use two Cyphers' Incubators (Fig. 17), each holding from three hundred and sixty to three hundred and eighty eggs. The best position for them, where they will be free from draughts and yet get good air, is marked on the floor plan of the incubator room. Very explicit instructions are sent with each machine, and with a little care and ordinary ability they are easily worked out

FIG. 17

with satisfactory results. "Elaine," or "Argo" oil are the only oils that are fit to use in either incubator or brooders. Sometimes I have been forced to use other oils that were almost as expensive; they have invariably given me trouble, the wicks getting charred and a low flame resulting, after a few hours' use. With either of the two oils mentioned a well trimmed lamp will burn twenty-four hours, without any perceptible difference in the flame and with no fear of explosion. A midnight explosion in the incubator house or brood house disturbs one's peace of mind, and even with cement floors may cause a fire. So, above all things, personally see that the best oil is used, and that the lamps are properly trimmed and the flame kept at the right height. I make it a rule to look at the lamps a few minutes after they are trimmed and re-lighted, and again before I turn in for the night.

Light the incubator and run it for a couple of days before putting in any eggs. This will give you confidence, for when you see that the regulator can be safely trusted to keep the temperature even, you will know that it is useless to be running to look at the thermometer every hour or so. I look at mine two or three times a day, never more, except when the chicks are coming out, and then only as a matter of curiosity, not necessity. With strong, fertile eggs, an even temperature throughout the hatch means healthy, well formed chicks; with the same eggs an uneven temperature will give a good many cripples, and the whole batch, with few exceptions, will start life under adverse conditions. When you see all the chickens with perfectly even, fluffy plumage, and of good size, then you know that the eggs and the hatching conditions were good. Otherwise, you will have a lot of sleek, uneven birds that look as if their fluff had been clipped with scissors; this is almost a sure sign that the heat has been uneven, causing the growth to be uneven also. The process of incubation should be steady and even all through, to obtain the best results.

The stock from which the eggs come is really the most important factor in raising chickens. From careful experiments that extend over a period of five years, I find that the eggs from pullets, i. e., birds under a year old, do not give satisfactory results, even when mated to a two year old cock. Nearly every pullet egg will hatch, but not fifty per cent of the chicks will live, and those that do live grow slowly, and reach the laying stage from two to three months later than birds that were hatched at the same time, but from fully matured stock.

This experience has been verified time after time; in fact, I have made a point of hatching from yards of stock that are matured and also from immature stock, keeping the resulting progeny separate, and noting results, so I am talking of something that I am sure of. A coincidence does not as a rule happen several consecutive times.

A hen that has moulted for the first time and is about fifteen or twenty months old, mated to a year old cock, is almost certain to produce, healthy chickens. I get from fifteen to twenty per cent less chickens hatched from the fertile eggs of this mating than from the same number of pullet eggs; but nearly all of the chickens live, grow fast, and begin to lay at the proper time.

In this climate I find that the best time to hatch the large breeds, such as the Asiatic and American varieties, is in February, March and April, and the small, or Mediterranean breeds, from March 1st until the end of May. If the White Leghorns are hatched before March, they will invariably moult the same year, which means that they will not be ready for steady laying before January or February; but if hatched in March, April or May, they will get down to steady laying at six months old and keep steadily at it until the following August. when they will come to their first moult. In other words, birds that are hatched too early moult the same year, and by the second August will have laid for about six months; whereas birds hatched at the right time will have laid for ten months before they come to their moult. We have here one of the principal reasons for the many failures in the chicken business.

It is not advisable to hatch after June 1st. The weather is then growing too hot for quick, healthy growth, and the vitality of the parent stock is then growing low. Repeated experiments have so fully convinced me on this point that I unmate all my yards on July first, and refuse to sell eggs for settings after that date until the following January. This means a serious financial loss to me, especially as so many breeders of good repute sell settings all through the year to customers who "have a hen that wants to set." These settings must be either from very immature pullets, or from hens that are moulting; in both cases the majority will certainly be weak. A bird that has laid heavily for nine or ten months and then goes into her moult, which is another heavy drain on her vitality, is in no condition to lay an egg that is fit for hatching. In the long run, I believe it pays to treat your customers as you treat yourself, and not to sell eggs for hatching purposes that you would not breed from yourself. This is another mile-stone on the road to permanent success, to know when to refuse to make a sale, even when the customer is willing and anxious to take all responsibility; for in time, when the eggs have hatched poor, weak chicks, the customer is apt to forget his readily assumed responsibility, and to lay the fault where it really belongs, namely, at the door of the seller.

Inbreeding is a bugbear that frightens many beginners. If one keeps several yards of the same strain of stock, for several years no fresh blood will be necessary; but when it is, be sure to try to get eggs, cockerels, or cocks from the original strain. I raise all my own pullets, and when I think that I want new blood, I get either settings or cockerels from the people from whom I bought the original stock. I therefore, in this way, keep the same strain, constantly improving it by careful selection.

Breeding from two different strains, even if both fowls be fine, thorough-bred specimens, often results in ordinary, mediocre progeny, especially if the breed is parti-colored, such as Barred Plymouth Rocks. It may take years of inbreeding to bring the birds up to the standard of the first parents. And fresh blood from a different strain sometimes has a disastrous result on the birds' laying qualities. If the new cock comes from a poor laying strain this quality is often transmitted to the offspring.

If after the first moult is over, one picks out the choicest hens for his breeding yards, sees that they have plenty of exercise to keep them from getting fat, and then in January mates them with some of his finest cockerels that look strong and mature, the chances are that the progeny will show the results of a good mating. This can be done for at least three or four years. Just watch the birds and their records carefully, and do not change, unless at the end of a season the number of eggs per hen has fallen below the average. I consider it impossible to lay down a set time for bringing in new blood, but I do believe that it is done a great deal oftener than is necessary, and sometimes with very poor results.

To summarize the above, as far as White Leghorns are concerned: Select the eggs for hatching from hens that were hatched between March and June, moulted the following year in the fall, and were mated in January or February to vigorous cockerels of from nine to twelve months old, the hens being from twenty to twenty-four months old. These ages for mating have given me the very best results. The birds should be mated at least ten days before the first eggs are gathered for hatching.

For our plant we shall want three hundred and eighty eggs every fifteen days, beginning from February 10th to April 18th. This will make the actual hatching from March 1st to May 1st, or a period of three months. The eggs can safely be gathered for the ten days prior to each hatch, but should be gently turned or rolled over every day. I have several shelves with wire bottoms fixed in the incubator room, and each day I select from the day's gathering about thirty-eight of the finest eggs, of even size and with glossy, white shells. Reject all small and over-large eggs, and any that are crooked, or have ridges on the shells. This last is a sign that the hens are too fat, and if hatched, the chickens from them would very probably be weaklings. Place the selected eggs on the wire shelves, and each day move them gently by passing the hand lightly over them. Some people stow the eggs carefully on their sides, with the large end slightly up, and turn each egg separately. This takes time, and is no better than if the eggs are allowed to lie loosely all over the shelf.

Let us suppose that the yards were mated about January 20th, and eggs began to be collected February 1st at the rate of thirty-eight a day. By the 10th there would be three hundred and eighty. If an incubator was lighted on the 8th, it should be well regulated by the 10th and ready for the eggs. The Cyphers machine begins the period of incubation with the thermometer at 102½ degrees F., and if the regulator has kept the heat at this point for twenty-four hours before the eggs are put in, and the operator has become familiar with the size of the lamp flame required, we can safely consider the machine ready for work.

From now on to the end of the hatch, I would advise you to follow strictly the instructions that come with each incubator; they are easily understood, and absolutely cover the ground. Do not deviate from them, and especially, that rule which strictly prohibits opening the door during the last twenty-four hours of the hatch, while the little chicks are coming out.

If all the foregoing directions have been carried out, at the end of twenty-one days there should be from two hundred to two hundred and fifty fine, strong, healthy youngsters ready for the brooder. When the hatch is over, open the door of the incubator and take out the trays with the empty shells, and eggs that have failed to hatch. Quickly free any chicks that happen to be almost out of the shell, put them back in the incubator, and close it again; but any eggs that are slightly pipped, leave alone as a bad case.

After the hatch is finished I keep the chicks in the incubator for twenty-four hours, with the thermometer at 103, and the ventilators full open. In case of an extra large hatch, where there would be danger from over-crowding, pick out the strongest looking chicks and put them in the brooder, until the balance left in the incubator is sufficient to fill it without crowding. Do not let them get chilled while doing this thinning out; if the door of the room is closed, and the air warm, there is no danger. During the twenty-four hours that the chicks are kept in the incubator they will pant a good deal, as if too warm; this panting does no harm, but I find it a good thing to open the incubator door for a minute or two, every two or three hours, to let them have a little fresh air.

Have a big bucket of water handy when you open the incubator to take out the trays of shells and unhatched eggs, and after releasing and returning to the incubator any chicks that are almost out of the shell, dump into the bucket everything on the tray. There are sure to be several eggs that are pipped and contain live chickens; but don't try to open the shells and release them—just let the water kill them quickly. If a bird is not out and free by the end of the twenty-first day, it will never amount to anything; it will probably linger a few days or weeks and then die in the brooder. This is the one hard part of keeping chickens in large quantities, and should be learned at the start. It never pays to keep weaklings. Lice always attack them before the strong ones, and they are liable to bring sickness into the flock at the slightest change of temperature; in fact, they are always a menace to the general health of the plant. It is a fallacy to suppose that with a little care they will live and grow big enough to be sold as broilers. Ninety-nine per cent of them will grow light, and never be fit for anything; so the kindest and easiest plan is to kill off all weaklings as soon as discovered.

Experience shows that this lesson of killing off the unfit is generally the very last one to be practiced. They are easily seen, but the kind, though mistaken, heart of the owner always wants to give them another chance. This natural impulse should be stopped from the beginning, before sickness develops which might easily spread through the whole flock.

CHAPTER VII.

—

THE BROODER

The brooder is a very important feature of the poultry plant, as it is the nursery where the baby chicks spend the first six weeks of their life; and if its conditions do not meet their needs, a large proportion of them will abandon the struggle with existence at the outset, while the survivors will not make as good a start as they should.

Unless a brooder is evenly heated and abundantly ventilated, it is impossible for the chicks to thrive; and so many of those on the market not having these qualifications, I propose to go into the subject quite fully before going on to describe the brood house and its work. A brooder can be heated with dead or stagnant warm air, or with live, moving warm air. I consider all brooders heated with hot water pipes or an inverted can coming through the middle of the floor and kept hot by a lamp under it, as belonging to the first class. The pipes or cans give off their heat, but the air has no circulation—or at the most, very little.

My first brooder, a standard make, was a four section one, and heated by hot water pipes. On lifting the lid at night, after the chicks had been in it a few hours, the air was pretty bad; and by morning it was very bad. I lost a large percentage of the chicks, and rightly laid it to the foul air of the brooder. On comparing my losses with those of others that I read of in various poultry papers, I found that I was doing fairly well, for brooder work, but when I compared them with chicks hatched from similar eggs, by hens, I found that I was really doing anything but well with my brooder chicks. I lost more than two brooder raised chicks to one hen raised chick, and the survivors were slower in growth and weaker. When they came to maturity, I could notice little or no difference between the two lots, except that those raised by hens began laying sooner than the others, and had laid more eggs at the end of the season.

About this time I saw in an eastern poultry journal, called "Farm Poultry," a long article by the Dr. Wood, whose valuable paper on feeding I have embodied in this book; he had noticed the weak points of the brooders then on the market, and had invented one of his own, and had kindly given the results of his experiment

to the public through "Farm Poultry," calling his brooder the "Up-
to-Date Farm Poultry Brooder." On reading it, I saw at once the
value of his idea, which was that the hot air should be kept moving
all the time. I have mislaid his paper with diagrams, so have here
given diagrams of my brooders, which I made on his principle, with
a few minor improvements suggested during my three year's use of
them.

A glance at the diagrams will show the simplicity and efficiency
of Dr. Wood's principle of heating. A sheet of heavy galvanized
iron is heated by a lamp placed under it; the outside cold air is
drawn in through the air space on each side of the brooder, warmed
on the hot plate, and then, drawing up the funnel, it strikes the
under part of the hover and spreads over the backs of the chicks,

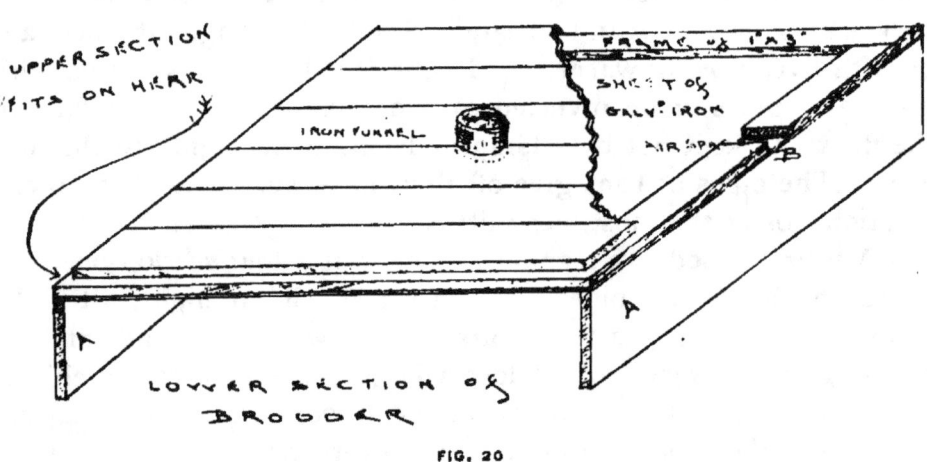

FIG. 20

after which it escapes from the hover and goes up through the ventilation slit at the top of the brooder. The result is that a steady
stream of warm, sweet air comes in and passes out all the time.

I made nine of these brooders without loss of time, and since
then, provided the eggs are all right and chicks properly incubated,
have had no further trouble with slow-growing, half-dead brooder
chicks! I shall try to explain the method of building them, for the
benefit of the amateur carpenter.

Take three pieces of surfaced redwood, two of them 1"x9"x3'
and one 1"x9"x4', and nail them together, making three sides of a
square (see A, Figure 20). Get a piece of No. 22 galvanized iron,
3'x4' and nail it on top of the frame A; then nail around the edge a
frame of 1"x3" surfaced pine, leaving an air space an inch and a half

wide at each side (see B, Fig. 20). Get some 1"x6" flooring—redwood is better than pine, as it does not warp with the heat—and
nail it over the frame, leaving one inch clearance **all** around for the
upper section to rest on. When the flooring is laid, cut a circular
hole in the middle of the floor nine inches in diameter; put a galvanized iron funnel into this hole one inch deep, and nail it from
the inside of the funnel to the edge of the flooring (Fig. 21), flush
with the under part of the boards. The funnel should be five
inches high above the level of the floor. Be sure to make tight
work of fitting the funnel, also of laying the floor, so that no sand
can drop through, and no heat escape except through the funnel.

Inside the funnel an iron cone is put, resting on the funnel
with three lugs, and with a clearance of half an inch. If the base

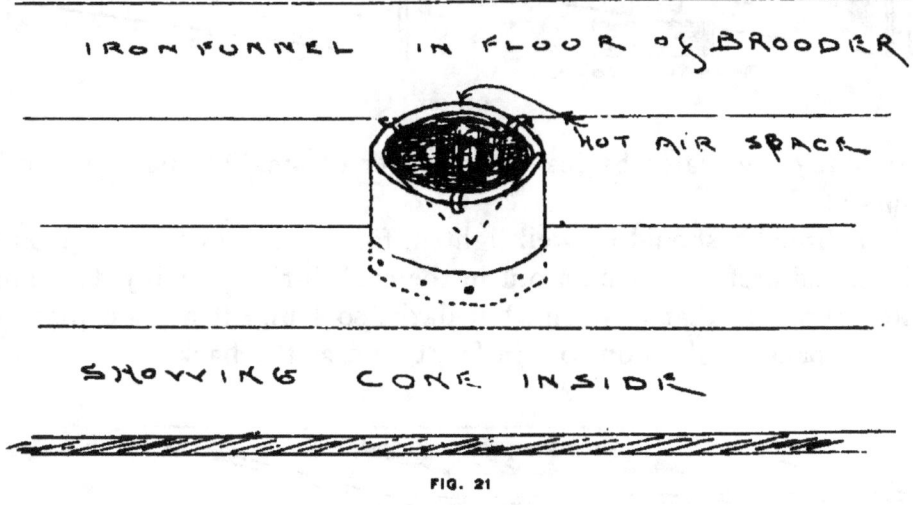

FIG. 21

of the cone is eight inches in diameter, this will allow the right
clearance. This cone is filled with gravel, which heats and keeps
the cone warm, thus allowing the warm air to come in contact with
a warm surface. The cone acts as a radiator, and it also prevents
the chicks from falling down on top of the iron plate when the
hover is off during the day.

The sides of the upper section can be cut from 1"x12" surfaced
redwood plank, and nailed to the flooring. The front (A, Fig. 19)
is made from a piece of 1"x9" redwood, with a hole cut in it to fit
the glass; this also is nailed to the sides and edge of the floor. The
top (B, Fig. 19) is better made from redwood flooring, which, being
tongued and grooved, will be tight, and prevent draughts. A piece

of glass is fitted in the middle of it.   The back (A, Fig. 18) has two side pieces nailed fast, B. B., and a framed glass C., which is fastened with a button on each side.   This is taken out for convenience in cleaning.

The lid (D, Fig. 18) lifts off; do not hinge it, as it makes an

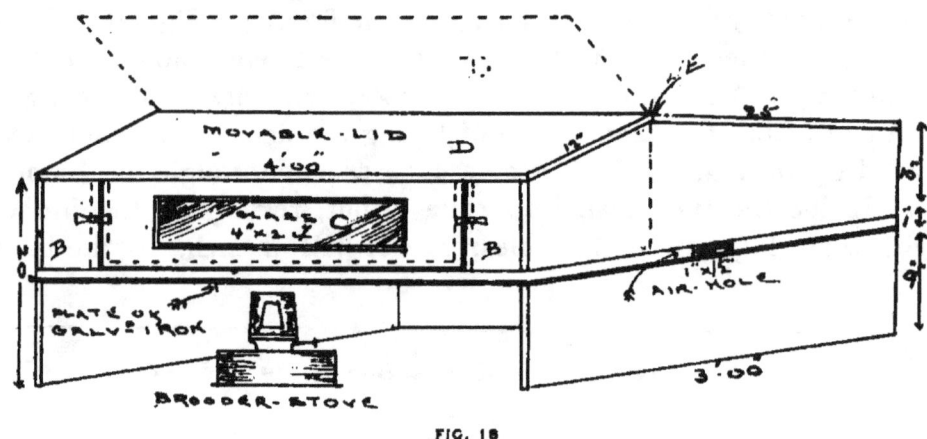

FIG. 18

adjustable ventilator by leaving a larger or smaller space where it joins at E.

A brooder should be well lighted, for the chicks shun a gloomy place, and prefer to remain out in the cold during the day, to going into a brooder that is warm, but dark; so I find it a good plan to have a pane of glass on top, in front, and at the back.   It means a

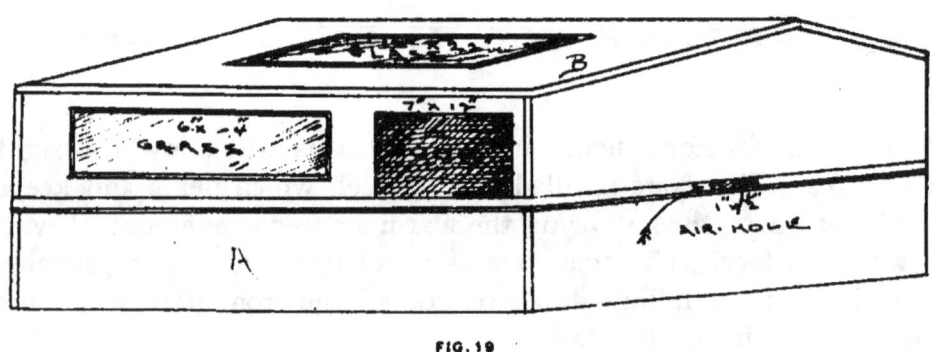

FIG. 19

trifle more flame, as glass is a better conductor of heat than wood is; but our chief idea is to have contented, happy chicks, even if it does take a little more oil.

At first I used for a hover a circular piece of wood, with strips of felt nailed to the edge, and coming to within half an inch of the

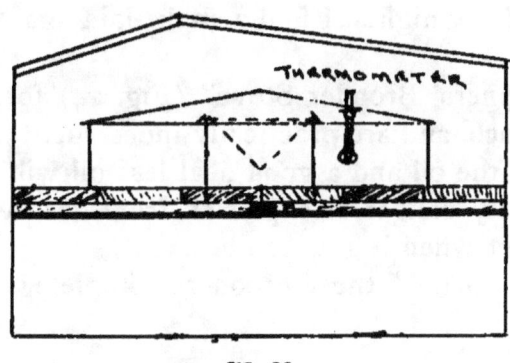

FIG. 22

FIG. 23

floor. I have discarded this for a better plan (Fig. 23), which is a circle of galvanized iron 28″ in diameter, and rising to a point in the middle, like a rather flat cone (B). From the rim a wall of iron comes down to the floor for one-third of the distance around, and opposite to it is a leg (C) for the open side to rest on. The wall rests in the sand, and goes in front of the entrance to the brooder, so that draughts strike the wall, behind which the chicks can lie snug and warm. The edge of the hover is 4½″ above the floor, and the peak (B) is 8″.

Before I made this style of hover I noticed that the chicks always got away from the side of the funnel near the door, for the felt flaps did not keep the draughts from going between and under them; but now they settle in a circle all around the funnel. This funnel should have a few turns of felt or flannel wrapped around it, as it gets too warm for the chicks to lie close to. Change this covering from time to time, as it will get dirty.

The hover has a hole in the top, through which a thermometer

FIG. 24

can be put for the first couple of weeks. For a door, which is only required the first few nights, I find a shake laid against the opening quite sufficient.

I use a "Cyphers' Brooder Stove" (Fig. 24) for each brooder; they cost $1.25 each, and are practically indestructible. Keep water in the pan under the oil and a good deal less oil will be burned, besides obviating any danger of explosion from gas, which generates in the oil chamber when it gets too hot.

When "home made," these brooders complete, with lamp, cost about $10 each.

CHAPTER VIII.

THE BROOD HOUSE

The brood house contains a total of twelve brooders, four of which will be required for each hatch. Each brooder holds from fifty to sixty chickens, according to the size of the hatch. I have raised eighty chickens successfully in a similar brooder up to the age of six weeks, but I do not recommend so many, as they are apt to crowd.

The floor of the brooder is covered half an inch deep with sand, or, if it is impossible to get sand, with soil. Each morning

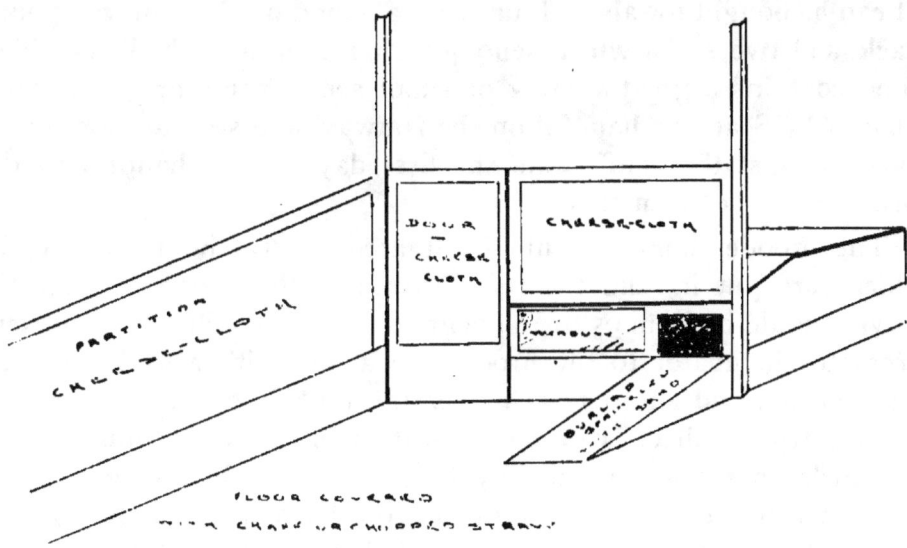

FIG. 26

the sand is screened, and once a week fresh sand is put in, after scraping the floor and spraying it lightly with Lee's Lice Killer. Do this in the morning, so that the fumes will have partly escaped before night. A small frame of 1"x3" pine with a bottom of ordinary screen wire will do for screening the dirty sand; this allows the sand to fall through, leaving the dirt in the screen.

The inside yard (Fig. 25) in front of the brooder is covered about an inch deep with chaff, or clean chopped straw, which can be renewed at the end of three weeks, the second lot lasting until the chicks are ready to leave the brooder.

Clean water should be given each morning when the brooders

are cleaned, and the water tins (Fig. 26) washed thoroughly with

FIG. 26

a round scrubbing brush; at the end of the week all these tins must be boiled. A small grit jar is kept in each inside yard, filled with chicken grit and fine granulated charcoal. Both of these can be obtained from Henry Alber's Poultry Supply House. The only safe charcoal to use is Willow charcoal, which contains less acid and tannin than the ordinary charcoal of commerce. For the first week I prefer to feed the chickens on Croley's "Dry Chicken Feed."

It can be bought for about four cents a pound by the hundred-pound sack, and two sacks will be enough for a thousand chickens. This food contains a great variety of small seeds, some grit, and some charcoal. Scatter a handful on the runway, and several more handfuls amongst the chaff; and, the first day only, a handful in the brooder, in the clean sand.

The brooder lamps should be lighted several hours before the chicks are put in, and the temperature of the brooder under the hover should be from 95 to 100 degrees. I usually shift the chickens from the incubator to the brooder in a box with a sack covering the bottom, and another sack over the top of the box, to prevent their getting chilled. I prefer to shift them in the morning, so that the little ones can have all day to get used to their new quarters and learn to eat. If the weather is cold keep them in the brooder all the first day, the handful of seeds being sufficient for the first day's food; if it is warm, sunny weather, go into the yard after a few hours, and tap on the runway with your finger, which will bring them scampering out to pick at your hand. You can then gradually coax them down into the straw, following your hand. All this is really not necessary, for if left alone they will come out a few at a time, and by afternoon nearly all will be scrambling and scratching amongst the straw for the seeds, and clustering around the water, drinking from the first. I advocate spending a few minutes amongst them several times a day, from the beginning; it makes them tame and used to you; otherwise, as soon as you go into the brood house they will flutter away in all directions, from fright. I must warn you against making any quick movements at any time, when near the birds, young or old; and don't let others do it, either; it frightens the birds; and when once properly frightened, they are always afraid when any one is near them.

In warm weather the lamps can be lowered during the day; and after the first day or two the hovers can be taken out at the morning cleaning and be put back about the middle of the afternoon. Do not clean the brooders too early, but let the morning get warm first; and oil and trim the lamps from 3 to 4 p. m., before the day grows cool; for the first week see that the chicks are all in the brooder by this time, and keep them in and draughts out, by putting a shake in front of the doorway, which can be removed in the morning to let them out.

Towards the end of the first week mix a little cracked wheat with the prepared dry feed, gradually increasing the wheat until the fourth week, when it will be all cracked wheat; then mix whole wheat with the cracked, so that at the end of the sixth week they will be eating all whole wheat. When the chicks are about ten days old, give a little of the following mixture every morning, but no more than they will eat up clean in ten or fifteen minutes. At first they will be dubious about it, but if left alone they will soon take it. As they grow older the quantity will have to be increased, but never give them so much that any is left on the plates an hour afterwards. With a scoop or measure thoroughly mix up the following ingredients in a sufficient quantity to last several days or weeks.

> 4 scoops bran,
> 1 scoop middlings,
> 1 scoop corn or feed meal,
> 1 scoop oatmeal,
> ¼ scoop meat meal,
> 1 teaspoon salt.

Mix with sweet or sour milk until it is crumbly but not wet, and feed at once before the meals sour. It is better to have it too dry than too wet. Feed it on small, shallow tin plates, putting three or four plates to each yard to prevent the chicks from crowding.

If this is fed in the morning, and at noon and evening, sufficient seed or wheat is thrown amongst the straw to keep them busy scratching all day, they will thrive, and grow big frames covered with firm flesh and little fat.

When I first started I used to spend a good deal of time feeding the chickens at fixed hours five or six times a day; in fact, it sometimes became a matter of coaxing to get them to eat; and if I happened to forget them, they simply "lay back and yelled." My log book shows a good deal of mortality in those days, under this coddling bill of fare; whereas now the work of feeding is a mere bagatelle, and the chicks are sturdy and busy all day looking for food which has to be looked for, before eaten.

After the first four or five days, if the weather is warm, open the little door and let them out to the soil yard, and then, in a day or two more, they can be let into the alfalfa yard, if the dew is off the leaves. The soil yard should be hoed and raked every two or three weeks; this keeps the soil sweet and crumbly, and induces the little chicks to take sun baths. It is very essential that the wee birds are kept from getting damp or wet; so before opening the alfalfa yards make sure that the alfalfa is fairly dry.

After the first week the temperature of the brooders may be gradually lowered and the use of the shake in front of the door be discontinued, until, by the end of the fourth or fifth week, the lamps can be put out and the chicks have a week or two of hardening before leaving the brooder. In warm weather, with the sun shining on the low, sloping roof of the brooder, the temperature of the brood-house will be high enough to allow the lamps to be out the greater part of the day, even with both doors and ventilators open. Keep a good thermometer hanging on the north wall of the brood-house, inside; this will act as a guide for keeping the right heat. On hot days, leave the tops of the brooders open enough to let them air, and still keep the chickens from flying out. At night, when going around for the last time, look through the brooder windows at the chicks and see if they are all right; if they are crowding together too much it means that they want more heat, but if they are all outside the hover it is because they are too warm; remember, though, that the temperature generally drops several degrees between ten p. m. and four a. m., and that it is better to have it a little too warm than too cold; for they can always find a cool place close to the sides of the brooder, if the ventilation is good, and then as the temperature drops, they will gradually get closer to the center of the hover and when the early chill comes they will all be under the hover. This is the beauty of having good big brooders— not too crowded—where several temperatures can be found.

After each batch of chickens, the brooder should be first scraped, then thoroughly washed out, and sprayed with Lee's Lice Killer, or some other good disinfectant; the inner yards cleaned and sprayed; and both brooders and yards whitewashed and the windows cleaned. I consider this an essential feature of the work, for after fifty or sixty chickens have been living in the brooder and yard for six weeks, even when the sand is sifted every morning and changed each week, there still is a pretty strong smell hanging around. If it is neglected, each successive lot of chicks will do worse, as the smells and dirt increase; and a stunted chicken never really grows as strong or lays as well as the one that gets no set-back, but grows straight along from the first.

When I hear about brooder chickens, or any chickens under five or six months of age that have been kept away from the grown hens and still have insects on them, I always consider it a grave reflection on the system of cleanliness and order practiced at that particular plant. After birds get matured they are very liable to get a few insects on them in this warm climate, even under the best conditions; but there should be no excuse for young chickens, hatched in an incubator and brought up in brooders, being infested with them.

CHAPTER IX.

THE CHICKEN HOUSE

By setting off an incubator every fifteen days, at the end of six weeks from the first hatch the twelve brooders will be full of chicks in three stages. There will be about two hundred at two weeks old, two hundred four weeks old, and two hundred six weeks old, with another lot nearly due. So it is time to shift the six weeks old birds to the chicken houses.

These houses should have sand on the floors, and for the first week or two, especially if the weather is cold or wet, put a couple of inches of clean straw over the sand, turning it over each morning and changing it at the end of a week. The straw makes the place cozy and warmer than the cold sand. The chickens, having been used to the wood floor of the brooder, have to be gradually acclimated to outside conditions.

The chicken house has three low perches, and the birds will come by degrees to roosting at night until, at the end of the second week in their new home, all, or nearly all will be up, and the straw will be no longer necessary.

Shift the chickens at night, putting the contents of one brooder in each house. The little openings in front of the houses have two slides each, one of wood and one of wire; for the first few nights

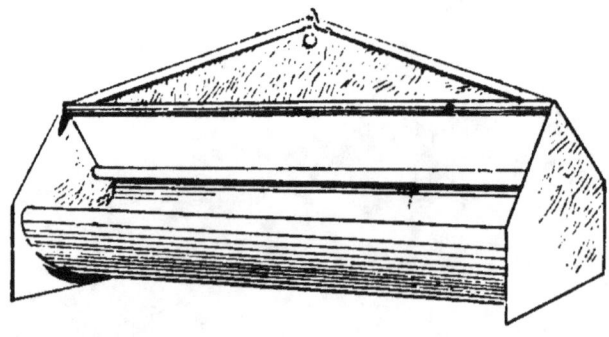

FIG. 27

close the opening with the wooden slide; afterwards, unless the night is very cold and wet, use the wire slide, until the birds are eight or nine weeks old, when no slide need be used unless the weather is rainy. While the birds are young it is important not to

let them get wet, so the wire slide keeps them in until you
see how the morning looks. If it is going to be a wet day, better
give them some straw with wheat or cracked corn scattered in,
to keep them scratching and lively; put in their drinking troughs
(Fig. 27), and keep them safely housed until the weather im-
proves.

On wet days I feed half cracked corn and half wheat; the corn
generates more heat than the wheat does, but if all corn is used
the change of food is apt to disagree with them, so it is better to
mix it with wheat, which they are used to.

Instead of using plates for the mash, as in the brooders, use
small wooden troughs (Fig. 28), which should be kept scrupu-

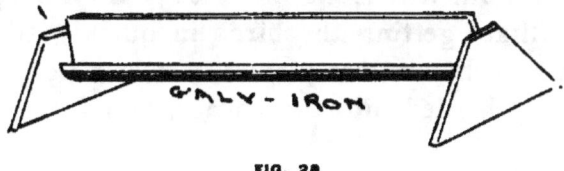

FIG. 28

lously clean. There are three sizes of grit sold, the small size for
the brooder chicks, an intermediate size, and a large size. For
these chickens use the intermediate size, mixed with granulated
charcoal. Keep a jar or two of it in the yard all the time; nothing
is better for keeping their digestive organs in good order than
plenty of grit and charcoal. If enough mash is given them in the
morning to satisfy them, with none left over, and some grain
always kept in a few inches of straw in one corner of the yard, the
alfalfa will do the rest of the feeding necessary.

The big alfalfa patch in each of these yards will be a source
of food and exercise, and keep them busy all day long. The importance of the alfalfa patch is only realized, by keeping a flock of
growing birds on bare ground, and another on alfalfa. The latter,
with one half the attention that is bestowed on the bare ground
lot, will in a given time be fully a third larger, and a good deal
healthier.

In fine weather, after cleaning the house, leave the hinged roof
open and let the sun play in all day, until evening, when it can be
closed at the same time that they get their grain. In the morning,
when feeding the mash, clean and refill the water tin. Time is
valuable on a poultry plant; therefore try to systematize the work,
and save both time and footsteps whenever and wherever possible.

For cleaning the sand in these houses, make a frame like the one used in the brooders, only larger, and with a coarser meshed wire for the bottom. Once a week sweep out all the sand, scraping off any dirt that may have stuck to the cement, and spray the floor with Lee's Lice Killer before putting on clean sand. I use a small hoe with the handle cut down to about nine inches, for scraping. It is well always to have this hoe handy when cleaning, to loosen any dirt that may have stuck to the perches or floor.

These houses are the only ones that have to be watched in case of the weather changing from fine to wet, during the day. If it looks like rain, run all the birds in and close the wire slide. A few drops, or a passing shower, will not hurt if the sun comes out warm afterwards; but if it looks like steady rain, keep them in the house. To facilitate getting the birds in quickly at the approach of rain, the chicken houses are placed in corners of the yards; the birds being easily herded into the corner, and from there into the houses.

Twelve weeks from the first hatch all the chicken houses will be full, as well as the brooders; altogether, there should be twelve hundred birds, and the hatching finished. Of this young stock, about six hundred will be cockerels that will have to be sold as "broilers" at the age of twelve weeks. On no account try to keep them until they are older, as the difference in price will little more than balance the extra cost of keeping them; besides, we now begin to need all the room for the pullets. So, when the first four flocks of fifty birds each are twelve weeks old, pick out the cockerels (they are easily known at this age by their longer legs and more highly developed combs) and sell them. The houses will then have only about twenty-five pullets in each; put the birds of two houses into one, then clean and wash out the two empty houses, whitewash them, and get them ready again for brooder chicks. The yards should also be swept, and clean straw put in the corners. When all is ready, take out half the birds from each of the brooders that have six weeks old chickens, selecting the strongest and best grown, and leave the rest to remain in the brooders until they are twelve weeks old. The hundred or so birds taken from the brooders are put fifty in each chicken house. This order of moving will have to be kept up until there is room in the big hen houses for the pullets, that is, until the old birds have been sold and the houses cleaned and whitewashed.

When birds are kept in the brood house over six weeks, the brooders should be closed, and small movable perches put in the inner yards for the birds to roost on, instead of going into the brooders at night.

The foregoing is the routine and order of housing after the first year, when the hatching commences with a stock on hand of a thousand hens, half of which will be sold in the early fall, so that only five or six hundred pullets will be required to take their places. The hatching for the first year will have to be just double as all the hen yards have to be filled, and some thousand pullets are required. For this we shall have to have ten hatches of two hundred chicks each, or five months of hatching, instead of three months. In this case I consider June or July hatching excusable; but under ordinary circumstances, do not hatch after May. The June and July birds will not get down to steady laying before February, which makes their laying season (to August) too short for the biggest profits.

The chickens hatched during the first year will, therefore, have plenty of room; and just as soon as they reach the age of twelve weeks, the pullets can be shifted to the hen houses, the cockerels sold, and the houses cleaned out and filled again from the brooders; and the brooders, in turn, washed, cleaned, and refilled from the incubators.

The mash for the six to twelve weeks old chickens is the same that the brooder chicks get; only put double the quantity of meat meal in it, and mix it with water if there is not milk enough for both. Always give the youngest ones the benefit of the milk, when there is not enough to go around.

When shifting the pullets to the hen yards, have a good pair of shears, and cut the feathers on one wing of each bird. At this age they are apt to fly like pigeons, and if put into the new yards without this precaution they will fly from yard to yard, and be a perfect nuisance; but if the wing is cut they soon get used to their new quarters, and by the time the feathers have grown again they are contented, and a good deal heavier, and less likely to fly.

This is the best time for culling the birds. Any that are deformed, or have crooked backs or no tails, should be sold or eaten, but on no account kept. Our layers are our profit makers, and must be in perfect physical condition, or we shall be feeding two cents' worth of food for one cent's worth of eggs. This, of course, does not apply to the plumage; a few colored feathers will make no difference to the laying.

I particularly wish to impress the importance of culling. "Keep no birds with physical defects if you want to succeed with poultry;" this is a rule that should be painted in big letters and stuck up in the brood house, and also at the entrance to the chicken houses, and, above all, over the portals of the hen houses. I firmly believe that every deformed or poor, weak hen not only leaves no profit, but also eats the profit of one good, laying hen. Beware of making your establishment a home for incurables. A little study of this doctrine will make one more successful, if harder hearted.

After all the chickens are out of the brood house and chicken yards, cut the alfalfa close to the ground with a grass-hook, or scythe, and keep it cut every month until just before hatching season opens the following year. The cut alfalfa can be fed fresh to the hen yards and will help out the lawn clippings. I use a "Mann's Clover Cutter," which cuts the alfalfa into short lengths, and if the alfalfa is cut while young and tender, the hens will eat every particle of it.

A doctor told me recently that if the first joint of one wing is cut off with a blunt pair of scissors, when the chicks are a day old, it will keep them from flying and not hurt them.

CHAPTER X.

THE HEN HOUSE

These houses are large and airy, and will comfortably hold seventy-five hens. Except during a driving rain from the south, the flaps, which face south, are opened when cleaning in the morning, and left open all day, until egg gathering time at night; this insures a dry, sweet house. At night sufficient ventilation is afforded by the space around the roof, and the ventilation space at the back of the house near the top. Many advocate bottom ventilation, others an open space in front with a curtain of cheesecloth or other light material. Some experts have considered my house far too small. My reply is that the birds kept in these houses are always healthy, lay well, and appear comfortable. My books show that I lose yearly from one to two per cent of grown birds, of which there are always at least a thousand on the place. With results like these, I consider it a good plan to let well enough alone. My theory is that in this climate, where the days are hot and the nights often twenty or thirty degrees colder, the birds require a fairly warm house at night, so that the outside difference in temperature is not felt inside the house.

I have no scratching houses, and consider them, in our climate, an unnecessary nuisance. The birds are always laying amongst the straw, and when one has to gather on an average five hundred eggs a day in the busy season, it is quite enough work to take them from the nest boxes, without having to rummage through a lot of scratching sheds—to say nothing of the risk of finding stale eggs; for every egg found amongst the scratching straw should be kept separate, and tested before being put with the market eggs. With two large yards, each 30x180 feet, to each house, and some trees or bushes for shade, the birds get all the exercise necessary, especially if the grain is scattered well over the yard and not thrown down in a heap.

Each yard raises a crop of wheat every year; some of this is eaten green, but a lot heads, and the birds jump up to get the ears and break down the stalks, covering the yard with a layer of heavy, coarse straw.

The ground being so rich with the continual droppings, raises enormous crops of high, coarse straw, which lies over the soil nearly all the year. This affords good, natural scratching, and being outside, seems to offer little or no inducement to the hens to steal their nests. Occasionally an egg is found, when scattering the grain, but not often.

When the twelve weeks old pullets are first put into these houses no nest boxes are required; they are therefore hung up to the rafters, out of the way. When they are needed, place them in position and put in each nest two or three inches of sand, and a nest egg. I prefer the medicated nest eggs. After the pullets have got used to the nests, these eggs can be taken out and kept for the following year.

I found that outside nests were not as convenient as nests in the house. With the former in use, the hens often laid in corners of the house, and cracked eggs were very frequent; but now it is a rare thing to get an egg outside of the nest. When gathering eggs, which is the last thing to do in the afternoon, any dirt or feathers found in the nest should be thrown out. A good plan is to empty the nests in two houses every week, on a given day, scrape off any dirt sticking to the bottom of the nests, spray them with Lee's Lice Killer, and fill up with fresh sand. In this way each nest is sure to be cleaned once in six weeks, and a habit is formed, which after a few times becomes natural and requires no further thought.

The mash for the pullets and hens is much the same as for the chickens, with the addition of more meat meal, and ground oats and barley instead of oatmeal.

> 4 scoops of bran,
> 1 scoop of middlings,
> 1 scoop of corn or feed meal,
> 1 scoop of ground oats (hull and all),
> 1 scoop of ground barley (hull and all),
> 1 scoop of meat meal,
> 1 ounce salt.

In the feed house are bins, each holding several sacks of the above articles, and one large bin, sufficient to hold about a ton of the dry mixture. I usually fill this bin with the mixture, putting in a bucket or big scoopful of each ingredient in turn, and then thoroughly mix with a hoe, until the bin is full. This quantity will last about a month and no time is lost in the morning; all that is necessary is to take so many buckets of the mixture and wet it with water in the mixing box, a box 4 feet long, 18 inches wide and 12 inches high. After the box is empty, tip it up in the sun to dry and sweeten, and once a week wash it out very clean. Be sure not to make this mash wet; have it just crumbly, no more.

The water tins should be placed in the shade, and be large enough to hold sufficient water for twenty-four hours. One tin to each yard is enough; when refilling, wash it out with a scrubbing brush, even if it looks clean. Fresh, cool water, and plenty of it, is most necessary, for over sixty per cent of the egg is water, and a shortage of water inevitably means a shortage of eggs.

The green food—either chopped fresh alfalfa or fresh clippings from the clover lawn—is better relished at noon than at any other time. I find that when I can feed plenty of green stuff, there are more and better flavored eggs, and a big saving on the mash and wheat. Besides, this bulky kind of food is necessary in keeping the digestive organs in good working condition.

The wheat is best fed at three p. m., scattering it well and giving a little more than can be eaten that night. Some grains are then left to be hunted for in the morning, keeping the birds busy and giving them a good appetite by mash time.

At the beginning of the rains, some cracked corn should be bought—not more than a ton at a time, as it deteriorates after being cracked or ground. On rainy days, mix half cracked corn and half wheat for the night feed, and if it is raining at feed time, instead of scattering it, put it into the feed trough, so that the birds can get it quickly and go back to the shelter of the house, out of the rain. During the rainy season I put a few drops of "Pacific Roup Cure" in the water, just enough to slightly color it. This preparation has a carbolic basis, and although I have never had the slightest approach to roup amongst the fowls, I believe in using a preventive during the season when colds are most prevalent. This carbolic acid makes the water look soapy, and care must be taken not to put too much in, or else the birds will refuse to drink it.

In emptying the water from the tins when cleaning them, scatter it, so that no puddles accumulate on the ground; the fowls curiously prefer these dirty puddles to the clean water in the tins.

I will give an outline of the average day's work in the hen yards, of which, on my place, there are twelve.

Seven a. m. The morning's ration of dry feed is put in the mixing trough and thoroughly mixed with water to a crumbly mass, which is then left to soak until feed time. The man then takes his big wheelbarrow, containing a bucket of clean sand, a small short-handled hoe, a short-handled corn broom, and a dust pan (Fig. 29) made of strong galvanized iron after the model used

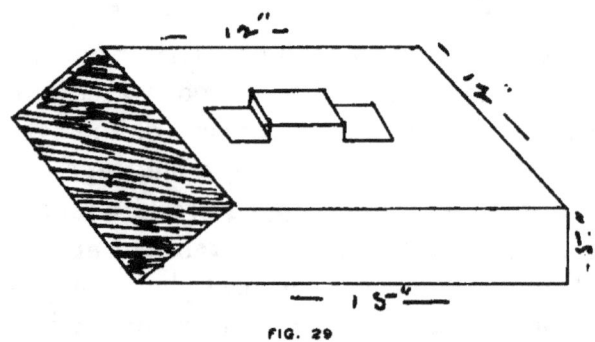

FIG. 29

by street sweepers; he goes to the first house, opens the flap to let the sun in, and then lightly sweeps the droppings under the perches into the dust pan, using the hoe where anything has stuck, either to the floor or perch. With practice, all the manure is taken, but very little sand. A few handfuls of sand are then sprinkled on the bare places on the cement floor. This is repeated in each house, and it takes from an hour to an hour and a half to do the twelve houses, according to the expertness of the cleaner.

The Leghorns are naturally high strung, nervous birds and great care must be exercised to avoid frightening them. Open the flap and door gently, and if any birds are still in the house, shoo them out without startling them. When sweeping, move quietly, so as not to disturb the birds on the nests. The lid of the nest boxes prevents them from seeing anything of the cleaner except his boots, for even different colored clothes, such as white trousers one day and dark the next, is sufficient to upset these nervous little creatures; so make the cleaner get into the habit of moving quietly, whenever near the yards. A man that persistently forgets this, and frightens the birds, should be dismissed in spite of other good qualities; for birds that are afraid of their care-taker will lay from twenty-five to fifty per cent less eggs than they otherwise would.

The sand box is conveniently placed for refilling the buckets, as three or four are required before the cleaning is finished. After the cleaning, the manure is screened free of the sand, which is wheelbarrowed to a dump, and the manure thrown into the compost house.

By 8:30 the cleaning is finished, dirty sand thrown away, and manure stored. The mash is now put into buckets and wheeled to the first yard, where so many scoops of it are put in each of the four troughs (Fig. 30). When in the yards, talk to the birds; the

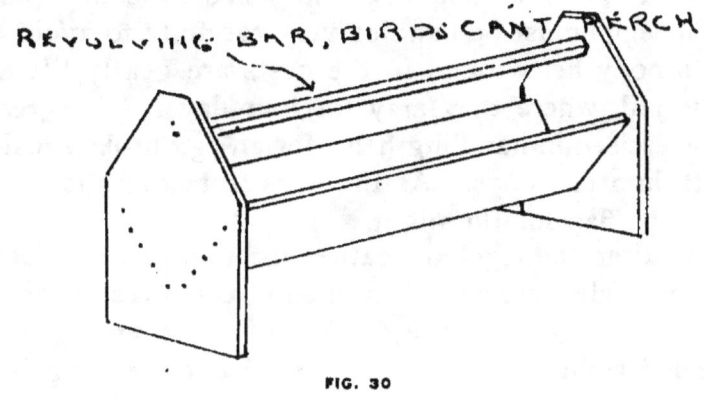

REVOLVING BAR, BIRDS CAN'T PERCH

FIG. 30

sound of the voice reassures them, especially at feeding time. Any idiotic sound will do, so long as it is not too loud, and bears a slight resemblance to their own crooning. I have often stood in front of the house, talking gibberish to them, and after a little while have had every hen answering me back. While feeding, keep your eye lifting to note how the birds look, and if there is a droopy one, pick her up quickly and put her alone somewhere.

By 9:30 the feeding is finished and the feed room swept out. Sufficient clover is then mown, or alfalfa cut with a grass hook, and run through the Mann Clover Cutter, and put in buckets ready for use at noon. Odd jobs are then done until eleven o'clock, when the birds get their clover, and the water tins are cleaned and re-filled. These two things are done at the same time, to avoid extra steps, the man entering the yard with a bucket of water and another of green stuff. This is a good time to gauge the right quantity of mash for the morning feed. If the birds appear very hungry, feed a little more mash next time; or if some still remains in the troughs, feed less. Some of the yards will require more, or less, than the others, at different times. Keen observation is the only true guide to the correct quantity to be fed, as it varies all the time.

It will now be noon, and time for lunch. In the afternoon odd jobs can be done, such as gardening, keeping the paths in order, etc., until 3 p. m., when the wheelbarrow is loaded with buckets of wheat and taken to the yards. So many scoops are well scattered in each yard, an eye being kept well open for stray eggs laid in the yards. After feeding, which, in order to get the wheat thoroughly scattered, takes about an hour, the eggs are collected in chip baskets holding not more than fifty or sixty eggs each. Any dirt or feathers found in the nests is taken out, and in the case of a badly broken egg, the soiled sand is replaced with clean. The eggs must be handled very carefully to avoid breakage, and any broody hens found in the nests are gently lifted out and put in the jail, where they stay for four days. The greatest care must be exercised in handling hens, for an egg broken inside means sure death in a few days. As the eggs from each house are gathered close the flap for the night.

By the time the eggs are gathered, brought to the feed house, the dirty ones cleaned, and all packed in cases ready for shipment, it will be about 5:30, and the day's work at an end. But I take a stroll around right after dinner, to see that everything is in order, and no hens are outside the yards, or roosting in the trees. A cold, damp night will give a left out hen a cold, which may quickly spread to the others, with serious results.

All this, of course, applies to the work of the twelve hen yards only. The chicken yards and brooders are extra, but are only in use for part of the year, whereas the hen yards are always full.

FIG. 31

In each house is a grit box (Fig. 31) that is always kept full of grit, shell, and willow charcoal, all of which are absolutely necessary to the general health of the birds.

The photograph, "Gathering eggs," shows a house with the flap open and the opening covered with wire netting on the inside, to keep the hens from coming out.

The "jail" is also shown close to the nest box from which I am gathering eggs; it is fastened to the end of the house and at the back a small opening, just large enough for a hen to get through, is cut into the main hen house, and a small door is hinged to the top of the opening, a wire is made fast to the bottom of the door and the end brought out to the front of the jail; after the birds have been confined for four days the door is opened by pulling the wire and the birds fly through the opening into their house. By this means we only handle the fowls once instead of twice.

Leghorns are called "non-setters," but I find that if they feel broody and are allowed to stay on the nest for one night, they are harder to break up than many of the so-called "setters." I therefore make it a rule to carefully put in jail any hens found on the nests when gathering eggs; after dinner I stroll around and make sure that all the nests are empty.

The nest boxes are conveniently arranged and can be got at without entering either the house or the yards, and the jail of each house is placed near the nests of that particular house. I have put the door of each house at the other end, to be as far away from the nests as possible.

The floor of the jail is a movable grating, that can be taken out once a week and washed; it also allows the cool air to come through to the hen, which helps to break up the fever much quicker than if the floor were air tight. When cleaning the gratings, each week, the droppings under the gratings are swept up at the same time.

This style of jail is my own idea and is both a convenience and a time saver.

CHAPTER XI.

———

THE YEAR'S WORK

I think a brief summary of the year's work will give an idea of the various phases of our poultry plant.

We will suppose that it is about the beginning of November. Six of the hen houses are full of hens just over the moult, and beginning to lay a few eggs. The other six houses are full of pullets, ranging from five to eight months old; some of these have been laying for the last six or eight weeks, and are now settling down to steady work. All of the brooders and ten of the chicken houses are empty. The last of the cockerels have been sold some time since, all except twenty-four of the oldest and finest, which are occupying the other two chicken houses, getting ready for the mating season.

The rains are due now, and as soon as two or three inches have fallen the twelve empty hen yards are ploughed and sowed to wheat. Use a quiet horse and a small plough, taking care not to bark the trees. After the ground is well ploughed, sow the wheat rather thick and cultivate it in with a spring toothed cultivator. I prefer this to a harrow, as it softens the ground to a greater depth than if a harrow were used. The ground near the houses can be done with a spade and rake; this leaves a neater job.

It will take six to eight days to do this work in a thorough manner, and it often pays to have it done by an outside man, so that no time and precious rains are lost.

I have experimented with oats, barley, millet, rye and wheat, and found that wheat has the strongest tap root; the fowls can pick and pull at it without destroying it, as it is impossible for them to pull it up. With the other cereals, they no sooner pull at the stalk than up comes the root; they then drop it and pull up another and so on, until, in a week or two, it is all pulled up and lying dry on the ground.

At this time all the brooders and chicken houses can be washed out and thoroughly cleaned, then whitewashed. This is good work for rainy days, as it is all under cover. Before I forget, let me say that after the last chickens are taken out of the brooders and chicken houses, do not leave the dirt and sand lying around in them until the rains come, but sweep them out and scrape off every particle of dirt from floor and perches; they can then be closed until the time comes for cleaning and whitewashing.

Several days before the whitewashing, I get some leaves of the common cactus, chop them up and fill an ordinary kerosene barrel about a quarter full with them, filling up with water. A barrel of the best lime is bought, and until finished is always kept in a dry place, with some empty sacks stuffed in it to keep out the moisture. Some empty coal oil cans, with the heads cut out and strong wire handles put in, make good buckets; whitewash spoils iron, so I do not use the good galvanized feed and water buckets for this purpose. The night before I am going to whitewash, a couple of buckets are half filled with lime, and then nearly filled up with water; in the morning the lime is found slacked, and in fine condition for mixing. Take some of it in the bucket you are going to use, add a tablespoonful of salt and enough water to make a thin paste; stir it well until free from lumps, and then fill up to the right consistency with the cactus water, which will be thick and gluey, and takes the place of sizing, preventing the wash from rubbing off; the salt makes the wash a pure white.

I have used a pump to spray the whitewash on, but find a large brush much better for this work. The wash had to be strained through a fine mesh before it could be used in a pump, and after the job was finished a good deal of wash was found where it was not wanted—on the green paint—and had to be washed off; so the pump was condemned.

In January the cockerels can be mated with the hens, putting five or six in each of the finest yards. Towards the end of this month the yards that were sown to wheat will be ready for use, the wheat being two or three feet high. At night shift the feed troughs and water tins into the fresh yards and open the small doors leading into them, closing the others. In the morning the birds will swarm into their new pastures, and in a few days quite a difference will be noticed in the egg production. The empty yards should now be ploughed and sown as quickly as possible, to give the new crop every chance while the rains last. Towards the middle of February all the cleaning, whitewashing and odds and ends should be finished and the place in first-class order, before the busy season commences; for everything that can be done should be finished before March.

Commence selecting eggs for hatching about February 1st,

and on the 7th set off the first incubator, so that the first hatch will come off on March 1st. Arrange to have a hatch due on the first and fifteenth of each month until May 1st. If for any reason one or two poor hatches are experienced, it will be necessary to continue hatching until the fifteenth of May or even the first of June, but on no account hatch later; better sell fewer hens that year, so as to make up for the smaller supply of pullets.

From the first hatch on, the work will gradually increase until the end of May, when everything will be at its busiest. Every brooder, chicken house and hen house will be full. In addition to the thousand hens, there will be twelve hundred chickens to care for and feed. No holidays or Sundays off should be taken while there is young stock on hand; it is better to wait until later, for besides being the busy season, it is also the time when most money is made. In addition to the eggs, there should be seven or eight dozen young cockerels to dispose of each fortnight from the first of May until July. Fortunately the days are long at this time, so that work can be started at six o'clock instead of seven, and continued when necessary, after dinner.

Towards the end of July the eggs will have fallen off very considerably, and it will be time to arrange a market for the sale of the five hundred hens. Try to sell them at the rate of a hundred a week, and before they begin to moult in earnest. As each house is emptied, give it a first class wash and whitewash, and sweep away all dirt from around the house, and have things in order for the new pullets. The nest boxes will not be wanted for some time, and can be hung up to the rafters, out of the way. As soon as each house is cleaned, fill it up with young pullets.

By August, the five hundred hens that are to be kept over for a second season will begin to stop laying, and from now on through their moult will have to be carefully watched. The moulting season is a precarious time for birds, unless they are well cared for. From the beginning of August until the end of the moult, when they are laying again, I mix in all the mash a tonic called "Myers' Royal Poultry Spice." It is good for the pullets as well, and gets them into good condition for laying. I have tried several different kinds of condiments, but get the best results from the above. If the birds are kept in good health by being properly housed, yarded and fed, no condiment or tonic should be necessary, except at the moulting period for hens, and perhaps for pullets just before they come to laying age.

I consider nothing is more to be avoided than to be always doctoring and dosing the birds. If one of the flock looks droopy, separate it from the others for a few days and give it a little extra care. Mix a little of the Poultry Spice in its food, and put a few drops of the Pacific Roup Cure in its water. If this does not improve it, "off with its head," and bury it deep. I mentioned before that in rainy weather I use a few drops of this roup cure in the water, twice a week, as a preventive for colds. If a bird looks really sick, do not hesitate to kill it at once and bury it deep. It is better to kill a hundred than to have a thousand sick ones. Again, if a hen has been pretty sick, and you nurse her through, be sure not to breed from her, for this tendency to catch cold can be transmitted to the next generation.

By October, all the young stock should be housed in the hen houses, except a couple of dozen of the early hatched and choicest cockerels, that stay in two of the chicken houses until fully grown and ready for mating in January.

Now the heavy work is about over for the season. The pullets will be thinking of laying, and the hens will have most of their new clothes on. As soon as the moult is finished, you will notice that the combs of the hens grow very small and pale, and remain so for nearly a month. Not an egg will they lay while in this condition, although during the heavy part of the moult they give a few, even when they are almost bare of feathers. I suppose that at the finish all their vitality is drained, and they have to recuperate, and hence want a nourishing tonic—not a stimulant—to encourage eggs.

By this time the owner of the plant will probably begin to feel as if he, also, wanted a tonic and a rest for a few weeks.

CHAPTER XII.

———

SUPPLIES

A few words on the subject of food will not be amiss. For convenience in reference, I will again enumerate the mixed mashes that I use for the birds.

Brooder Chickens, From One to Six Weeks.

4 parts (by measure) bran,
1 part oatmeal.
1 part middlings,
1 part feed or corn meal,
¼ part meat meal,
1 teaspoon salt.

Chicken House Chickens, From Six to Twelve Weeks.

4 parts bran,
1 part ground oats (husks and all),
1 part ground barley (husks and all),
1 part middlings,
1 part feed or corn meal,
½ part meat meal,
1 ounce salt.

Pullets and Hens, From Twelve Weeks Up.

4 parts bran,
1 part ground oats,
1 part ground barley,
1 part middlings,
1 part feed or corn meal,
1 part meat meal,
1 ounce salt.

I mix the mash for the chickens with milk, and would use it for all the mashes if I had it, or were situated near a dairy; but as it is, I have to use water when the milk fails.

Do not mix over night, as the mash sours, and is not good for the birds; mix the first thing in the morning, and it will then get an hour's soaking before being fed.

The wheat I buy in August, or as soon as the thrashing is finished. I get twenty-five tons then, which just lasts the year.

Usually it costs from $22 to $25 laid on the place. It is not necessary to buy what is called milling wheat, or shipping wheat, which is clean and plump, and free from other grains; in fact, I do not care for plump wheat, as it contains more starch than is necessary for the birds. I try to get small parched wheat, that has five or ten per cent of oats or barley mixed with it, but is perfectly sweet and sound; it costs considerably less than milling wheat. On no account buy smutty or unsound wheat at any price, unless you want sick hens.

The store house is large enough to permit the wheat to be stored in several piles, which enables one to examine it occasionally, to see that it is not sweating. I always put some pieces of old board under the sacks, as cement sometimes gets a little damp when the air is cut off from it.

The bran, middlings, feed meal, ground oats, ground barley and meat meal are better bought a ton each at one time. This especially applies to feed meal, which deteriorates after being ground a month or so. The usual price for these articles is as follows:

Bran, $20 to $24 per ton.
Middlings, $25 to $35 per ton.
Feed meal, $35 to $40 per ton.
Ground oats, $30 to $35 per ton.
Ground barley, $22 to $26 per ton.
Meat meal, $35 to $40 per ton.

All of these except the last are bought direct at the mill; the meat meal I buy from the "Agricultural Chemical Works," 901 Macy Street, Los Angeles. This meat meal is tankage, especially prepared for poultry and contains cooked blood, bones and meat ground fine. I used to use fresh ground bones which I got from the butcher and ground myself in a "Mann's Bone Cutter," run by a small gasoline engine. I fed two hundred birds for a year with the freshly ground bone, and another two hundred with meat meal. Careful count was kept of the eggs from the two flocks, and at the end of the experiment there was a difference of twenty-three eggs in favor of the ground bone! I felt a trifle sad to think of all the extra and disagreeable work I had done for those miserable twenty-three eggs, and promptly discarded fresh bones, in spite of the many glowing reports of other people's results. It is a fine thing to keep a careful and reliable record.

The "Dry Chick Feed" is a mixture of many seeds with grit and charcoal; I prefer the kind sold by G. H. Croley, 508 Sacramento Street, San Francisco. It costs $4 a hundred, and two hundred pounds is sufficient, with the help of cracked wheat, for the season's hatching.

Grit, shell, willow charcoal, and everything else in the way of supplies I get from Henry Albers, 315 South Main Street, Los Angeles. I keep his catalogue handy; anything that I require, not mentioned in the catalogue, he gets for me.

Cotton seed meal costs about $40 a ton. During the moult, I use in the mash, instead of one part meat meal, ½ part meat meal and ½ part cotton seed meal, as it has a lot of oil in it, and in my opinion is a great help to the moult.

The tonic that I use during the moult is called "Myers' Royal Poultry Spice;" I feed it according to the directions, and buy it in lots of one hundred pounds at a time. It is a mixture of ground herbs, and, with me, has proved the best tonic I have ever used. I get it from the agents, "The Keystone Milling Co.," San Pedro Street, Los Angeles.

When the hay is cut I buy two or three tons of a good quality, and store it in the store house, ready for use during the hatching season. Egg cases, incubators, water tins, grit tins, "Union Lock" wire, in fact, nearly all of the fixtures required, can be got through H. Albers. I am afraid this looks like a puff for Mr. Albers, but it is not, in the way of business; simply a recognition of good service rendered, at fair prices.

One should take a paper or two, just to see what others are doing. Often they contain nothing that is particularly interesting to the commercial poultry man, but sometimes one comes across an article that is worth the year's subscription price several times over. For instance, Dr. Wood's brooder article in FARM POULTRY was worth paying the subscription to that paper for the rest of my life, even if I never got another idea from it.

A thing I object to in many of our poultry papers, and that one must be on guard against, is the way articles and letters are allowed insertion without being first edited by a competent poultryman; a lot of misleading nonsense is written that in many cases is accepted and acted on, as sound knowledge, by the novice.

The principal California papers that may prove interesting and of value to the rancher are as follows:

The "CALIFORNIA CULTIVATOR" (weekly). This contains departments for poultry, garden, orchard, stock, etc.; also the "Produce Market" quotations. I find this paper invaluable to one living in the country.

The "LIVE STOCK TRIBUNE" (monthly). Strictly a stock paper, in which are articles from the principal breeders of fancy stock in this state.

The "RURAL CALIFORNIAN" (monthly). This is principally devoted to horticulture, irrigation, and kindred subjects.

The "FANCIER'S MONTHLY." The oldest paper of its kind in the state, and, I believe, an authority in the fancy trade.

"FARM POULTRY" (semi-monthly), Boston, Mass. This is an eastern paper, and by some, might be considered out of place here. I like it for several reasons, the principal one being that the editor is a practical poultryman, and carefully edits pretty well everything that goes into the paper. He is severely practical, and ruthlessly expresses his opinion on everything that savors too much of the optimistic amateur. He is the author of "Poultry-Craft," which is in my estimation the standard work of its class, in this or any other country. Although I do not consider myself quite an amateur in commercial poultry, and think that I know more about our local conditions than Mr. Robinson, of "FARM POULTRY," I am quite prepared to accept in the right spirit any criticism that he may see fit to write on this book of mine, for I know that he will be just, if severe.

"THE FARM POULTRY DOCTOR," by the above publishing company, is a good book to have on hand in case of sickness amongst the flock.

"DISEASES OF POULTRY," by Dr. Salmon, is also an authority on the subject.

CHAPTER XIII.

FOOD VALUES

The chicken raiser generally passes through a "balanced ration" period and at the end of it, in his inmost heart, feels that "he don't know where he are." When I was recovering from my dose of it, which lasted for over a year, I happened to see the following article by Doctor Woods in "Farm Poultry," and felt so impressed by its strong common sense, that I take the liberty to reproduce it for the benefit of others.

FOWLS AND FOOD

THE LIVING FOWL--ITS FOOD AND WHAT IT DOES WITH IT--A LITTLE PHYSIOLOGY BOILED DOWN FOR POULTRYMEN

The living fowl is often likened to a machine, but the comparison is hardly just to such a wonderful complex organism as the living body. It is more like a living city peopled by a myriad of living cells, each with its duty to perform. There is a great supply system for receiving food and fuel, the digestive and respiratory organs, which with the tissues also represent great manufacturing plants capable of converting food and fuel into heat, work and building materials for the repair, maintenance or development of various parts of the body. With these manufacturing plants are intimately connected great store-houses to be called on in time of need, like fat tissue and the liver. There is a great system of transportation, the circulatory system, for carrying supplies and some workers to various parts of the body, and returning waste products to the excretory organs. Then the nervous system which has its sub-stations and telegraph lines communicating with all parts of the body, and which governs, polices, and regulates the whole. Presiding over all is something supreme, and about which we know next to nothing—life.

It is not strange that in attempting to convert this wonderful living body—the like of which we have no power to create, but which possesses the ability to reproduce itself—into a machine we meet with obstacles which we fail to understand the meaning of. The wonder is that we succeed in controlling it and making it serve us as well as we do.

The body is made up of an infinite number of living cells and their productions. These cells have varied duties to perform, and while some are confined to their special department and are gifted only with passive movement, there are others more active which travel all over the body. All are concerned in the maintenance of the body. Some of the active cells act as an army to repel invaders which appear in the form of disease germs. A few cells may neglect their duty and no harmful result is apparent, but let a number of cells combine, like the strikers in organized labor, and there is trouble until the dissatisfied population is put to rights again.

Chemically the body is made up of water, protein, fats, mineral matter and some carbohydrates (starches and sugars, these appearing chiefly as stored fuel manufactured from food). Accurate knowledge of the chemical compounds which exist in the living body and their exact disposition and relation to each other is impossible, as in order to make an analysis the complex living matter must be killed and broken down, leaving only the debris for examination. Accurate knowledge of the changes which take place during the digestion and assimilation of the food is likewise impossible, as we must first kill the fowl or induce an unnatural condition, before it is possible to observe what is going on within it. Obviously much must be left to be drawn from theory. The theories, however, are ably supported by the result of careful experiments based upon them. It is possible to prepare food of known chemical composition, and after feeding the same and making analysis of the waste disposed of, to estimate the amount of each constituent of the food digested. But the conditions governing the experiment are necessarily artificial, and the results do not show how the fowl disposes of what it digests or that a fowl would digest a like proportion of the food under normal conditions. The fact that the experimenter is dealing with a complex living organism subject to influences of which he has little or no accurate knowledge make it difficult to approach anything like exactitude in results. A skillful experimenter may "prove" almost anything he sets out to demonstrate to his own satisfaction when handling live stock. He can show that under certain conditions, with certain fowls and certain methods he obtained such results, but another man with different fowls may follow his lead as exactly as it is possible for him to do, and the results will be widely different.

It has been demonstrated by experiment with animals that the difference in individuals in the proportion of the food digested of a given ration is not as great as is popularly supposed. Several individuals of the same variety might "digest" a like amount of the different constituents of a ration, but the disposal each would make of the digested matter would vary widely. One might make heat or fat of it where another converted it into eggs or meat. The disposal of the digested matter will vary from time to time, "nature" choosing as she elects, and an exactly balanced ration that will cover all conditions and meet all individuals on a common ground from day to day is an impossibility. To balance a ration as accurately as some folks would have us believe they do it, would require the gift of second sight and a daily change in the nutritive ratio of the ration.

Such finickyness is not necessary or desirable. The chemical analysis of a food is of value only as it shows us the make up of the food, and saves us from feeding an excess of costly unnecessary material. In the majority of cases we cannot have an analysis of each lot of food purchased (even if desirable) and are obliged to depend on the average chemical composition for that particular food stuff. A glance over analysis tables of U. S. government reports will show that even the grains vary widely with the different samples of the same grain examined. As a rule it is safe to accept the whole grains at the average nutritive ratio set for them. With ground foods and meat foods, when trouble appears in a flock from unaccountable causes, it will be well to look carefully into the composition of these manufactured food stuffs and see if therein can be found a cause for the trouble.

The man who spends his time figuring out a chemically balanced ration, and wearies his brain with nutritive ratios and potential energies will not, in nine cases out of ten, be anything like as successful as the man who studies his fowls and feeds them according to their appetites on good plain food in variety.

It is essential that food shall be pure, palatable and digestible. A food may show an "ideal" chemical composition and yet be neither palatable or digestible. So far as the daily balance of the ration is concerned it is safer to leave that to the instinct of the fowl than to man's invention. The fowl's appetite is not an infallible guide, but if the fowl be given a fair chance to select its own food it can be depended upon to do fully as well as when it is provided with an elaborate man-concocted mixture.

Chemists and physiologists have demonstrated some things about feeding that appeal to common sense. It has been shown that the body is made up chiefly of water, protein, mineral matter and fats. The carbohydrates appear also, but chiefly as stored material (fuel). We know also that foods contain these compounds.

Water is supplied in the food and as drink. We don't have to pay further attention to that than to make sure that the fowls always have a supply of pure water for drinking purposes.

Protein is the most valuable constituent of food. Animal protein (contained in meat food) is considered more available and more perfectly digestible than vegetable protein. It has been shown that some animal food is necessary to health. How it differs from vegetable protein we do not know, but vegetable matter will not completely take the place of animal matter. Protein besides being the most valuable is the most costly. It also has the widest range of uses within the body. Its chief value is as a tissue builder. It furnishes material for tissue building and repair, and contributes largely to the manufacture of eggs. It is also convertible into fat and heat. The waste from protein is more dangerous and more difficult to get rid of than that of other food constituents, so that aside from an economical point of view, it is unwise to feed a very narrow (excessively nitrogenous) ration.

Carbohydrates are chiefly heat producing. They supply fuel, energy, to be converted into work and heat. It is doubtful if the carbohydrates are available for any other purpose. It is thought that they cannot be converted into fat, but act rather as a fat saver by furnishing fuel to be consumed in place of fat. Ducks fed abundantly on rice, which contains much carbohydrate and little protein or fat, remain lean; if fat is added to the food they lay on fat. The liver, besides manufacturing bile for use in digesting and assimilating food, seems to act as a manufactory and storehouse of partially converted carbohydrates, and it deals them out in the form of a starch that is readily convertible into a sugar easily assimilated by the tissues. Where carbohydrates are greatly in excess, a too starchy diet, the liver is overtaxed, and we get so called "liver troubles."

The fats are available for energy, for work and heat, and may be stored for future use, or so disposed of as to be of service as insulators to protect the body against too rapid loss of heat. They serve as fuel for growing and working cells. The fats are carried to the cells in the form of minute fat droplets and undergo chemical changes within the cells before being deposited in storage as fat tissue. The fats also contribute to egg formation.

Mineral matter is necessary to supply the tissues, form bone, and supply mineral matter for eggs and material for egg shell.

Oxygen from the air inhaled is taken up by the blood when passing through the lungs, and is carried to the tissues to help in the chemical changes which occur there.

There are other things necessary to properly supply the living body. There is something that is not contained in dry grain or meat food especially if the food be cooked. Live fresh green food is necessary to the health of the fowl. There is something contained in the live cells of fresh green stuff that possesses health giving, disease preventing properties. We do not know what that something is, but we know that it is there and that it is necessary. Cooking the food destroys the live cells and does not add anything to the food except bulk, and renders the starch more easily digestible. The cooking is chiefly of service in adding variety to the food and in destroying any undesirable germs which may be present in the meat food.

The true digestion of the food does not take place in the crop, stomach, gizzard, and intestines alone. It takes place all over the body, in the tissues. Suppose the bird to have been fasting. Food is taken into the crop, and the activity of that organ in supplying fluids to soften the food at once starts heat generation. The muscular contractions used to force the food onward also make heat. Heat production increases rapidly as the work of digestion progresses. After the food is reduced to paste by grinding in the gizzard and mixing with the secretions of the stomach, the intestines, the liver and pancreas, it is taken up by the circulation and carried all over the body to the tissues. There it undergoes chemical transformation, and its potential energy is further converted into kinetic energy in the form of work and heat. That portion of the food not available for the needs of the fowl, together with the waste brought back from the tissues by the circulatory system, is voided as droppings. The maximum of heat production, which began with the taking of food into the body, occurs some six or eight hours after the meal. The activity of the organs of the body, muscular activity, building up and breaking down of the tissues, all contribute their share to heat production.

Heat is lost to the body in a variety of ways. Some is carried off in the exhaled breath and in the droppings, and some disposed of by radiation from the surface of the body. Too rapid loss of heat is provided against by insulation of the body with fat and by clothing in the shape of feathers. Throughout the life of a healthy fowl this heat expenditure is under the control of a delicate system of regulation, a part of the nervous system. These heat regulator nerves control the rapidity with which heat is expended, and have power to excite heat production and so maintain the bodily temperature at a proper degree. Whether the heat production within the body be rapid or slow, the body temperature remains about the same all the time controlled by the regulator. The above only covers the subject roughly, but probably as fully as the average reader will find patience to consider.

The fowl, if permitted to range and find its own food, will live chiefly on grains and seeds, will drink freely of water, eat quantities of green food when available, and consume a considerable proportion of animal food in the shape of bugs and worms and any waste meat it can find. The nutritive ratio of such a ration must vary widely. Yet if the bird is on a large farm, has decent sleeping quarters and an occasional feed of corn on the ear, it usually does remarkably well, all things considered. Hen farmers who make a fair living out of poultry often let the fowls balance their own rations and keep boxes of corn, oats, and meat scrap always before the birds. Oftentimes these birds do quite as well as those fed in a "scientific" manner. Why? I don't know, but I think that the presiding life within the fowl, which dominates its nervous system and impels it to do certain things, is responsible for the success of the fowl left to its own inclinations. Where things do not go as they should some morbid condition has interfered with the normal conduct of the living organism.

So far as balancing a ration goes, I think that there has been and is a great deal of nonsense connected with it. Rations which vary widely in nutritive ratio are giving equally good results in the hands of different poultry keepers. It is undoubtedly wise to roughly balance a ration by offsetting a heavy supply of carbonaceous food with some nitrogenous matter, or vice versa. I know it is not necessary to provide elaborate mashes with a multitude of ingredients. So far as is possible it is undoubtedly the safest plan to observe the flock carefully, note the work done, and try to feed according to what seems to be the immediate need. Let the ap-

petite and inclination of the flock as a whole, combined with the
work it is doing, be the guide to the make-up of the ration. The
ratio may fluctuate from 1:3 to 1:9, with an average about 1:6, and
the results prove excellent. It isn't necessary to sit down and
figure it out. If you observe the flock carefully as a complete whole,
made up of individuals, and take notice of the effect of what you
feed, and the behavior of the birds, you will learn what foods and
feeding methods are best suited to your needs.

Supply good sound grains, some cooked mash by way of variety, a liberal supply of fresh green food (in summer the best way
to supply it is to provide a clean grass run), and some meat food.
Make grain the staple food and the others side dishes or relishes.
Avoid too much sameness in the daily feeding. Provide grit, shell
and charcoal for the fowls to eat as they please. Watch the condition of the fowls carefully, try to keep them well fed, active and in
good laying or breeding condition. The droppings should be of
sufficient consistency to hold their shape, but should not be too
solid. In color they should be dark tapering off into grayish and
white. If the droppings are watery and dark with red splashes of
mucus in them feed less meat food. If droppings are soft or pasty
and yellowish or brownish, feed more meat and less starchy food.
Greenish watery diarrhoea should always lead to a careful investigation of the sanitary conditions and the condition of the food and
water. It is a danger signal.

Exercise and fresh air are important to the proper assimilation
of food for best results from layers or breeders. Exercise prevents
misappropriation of food and the laying on of too much fat. A diet
that with exercise proves excellent for egg production will, if the
fowl be prevented from exercising, prove fattening. If you feed
a diet rich in protein, and neglect to provide for exercise, you simply pay for high priced food to produce fat which can be more
cheaply produced by feeding fatty food.

No rule for feeding can be given that will fit all conditions and
give like good results for all flocks if followed blindly. In an article like this the writer has no choice but to give the salient points
as well boiled down as possible. So far as the practical application
goes each reader must read, digest and apply the matter as fits his
own particular case and as it appeals to his common sense.

DR. WOODS.

CHAPTER XIV.

THE WATER SUPPLY.

If sufficient water can be obtained by digging a well, it will be cheaper, in the long run, to own your water supply than to buy it. If bought it is generally sold by meter measurement at from fifteen to twenty-five cents per thousand gallons; and when one uses two or three thousand gallons a day for eight or nine months of the year, it means a rather big bill.

If water has to be got by boring a well several hundred feet, and by no other way, I should prefer to look elsewhere for land. A bored well is expensive to bore and case, and is liable to get sand in it. Besides, a deep well means heavier machinery to pump the water.

If a dug well is decided on, care must be exercised when selecting the man to do it. Many wells are dug and cased in a manner simply disgraceful. I have seen some work done in well digging that cost twice as much as necessary, and the wells were poor affairs throughout; so see that the so-called "well digger" is not making his maiden attempt at your expense.

I have had two wells dug and set up with complete pumping outfits; and after five years' service, I am satisfied with them in every respect. I will describe the last one.

Before digging I inquired around to find out the man who had dug the most successful wells in the valley. Then I got him to go over my place and, taking into account the general contour of the country, and localities of other wells, to tell me where, in his opinion, was the most likely place to get water. Having settled this point, I arranged with him for digging my well, which was to be six feet square. He was to dig the well and lay the casing for $3 a foot for the first thirty feet, and $4 a foot for the next thirty feet. I was to supply the casing, which was 2"x12" redwood, costing, with corner pieces, about $1.25 for every foot cased.

The digging was pretty hard, being clay and boulders for twenty feet; many of the boulders were so big that they had to be blasted. My digger employed a man, boy and horse. After the first few feet, which could be thrown out, a small derrick was rigged over the hole, with a block and fall attached to it. A dirt bucket was made fast to one end of the fall and the horse at the other end,

7

which led through a lead block fastened to the foot of the derrick. When the bucket was full the horse, led by the boy, hoisted it, and the man on top tipped it on the dump pile. I found this system of hoisting far preferable to a hand windlass or a small hoisting engine, for a well of moderate depth.

As the well went down it was cased, so all dangers of a cave-in were avoided. At twenty feet water gravel was struck, which grew moister with increasing depth, until at twenty-five feet a foot of water would accumulate in an hour. By the time the thirty foot level was reached they were hoisting one bucket of dirt and two of water, so a common jig pump was installed and a boy put on to pump. This kept the well dry, and it was then found that the water gravel ended at thirty feet, and clay was encountered again. A sump was then sunk eight feet in the clay, and a platform made over it at the thirty foot level. The water then drained into the sump, which when full was pumped out. We found that we had fifteen hundred gallons in twenty-four hours, and as this was in August we reckoned that this would be our summer supply. (Always dig your well in the middle or towards the end of the dry season, to be sure of knowing what the lowest supply will be).

I went down with the well digger, and we made a careful examination of the water gravel and found that the water came from a northeasterly direction. I then decided to run a tunnel southeast through the gravel and cut the little water streams at right angles. The tunnel was started at the thirty foot level, where the gravel ended and the clay began, and was four feet wide by seven high, and well timbered from the entrance, to prevent caving. The timbers were 4"x6" redwood.

Frames 4x7 feet were made of these timbers, mitered at the joints, and placed every four feet in the tunnel. Behind and over the frames was put 2"x12" redwood casing. This made a solid sheathing on both sides and top. The level of the floor was uplifted from the entrance about one foot in fifty, which gave ample drainage. When the tunnel had been run in forty feet the fifteen hundred gallons a day had been increased to six thousand, which was all the water that we required. Lately, I went through the tunnel and found it in perfect order, not a timber started and the wood in good condition. There are several ways of casing a well, but I prefer this style (Fig. 32), as I think it not only the best, but the easiest to cut and place.

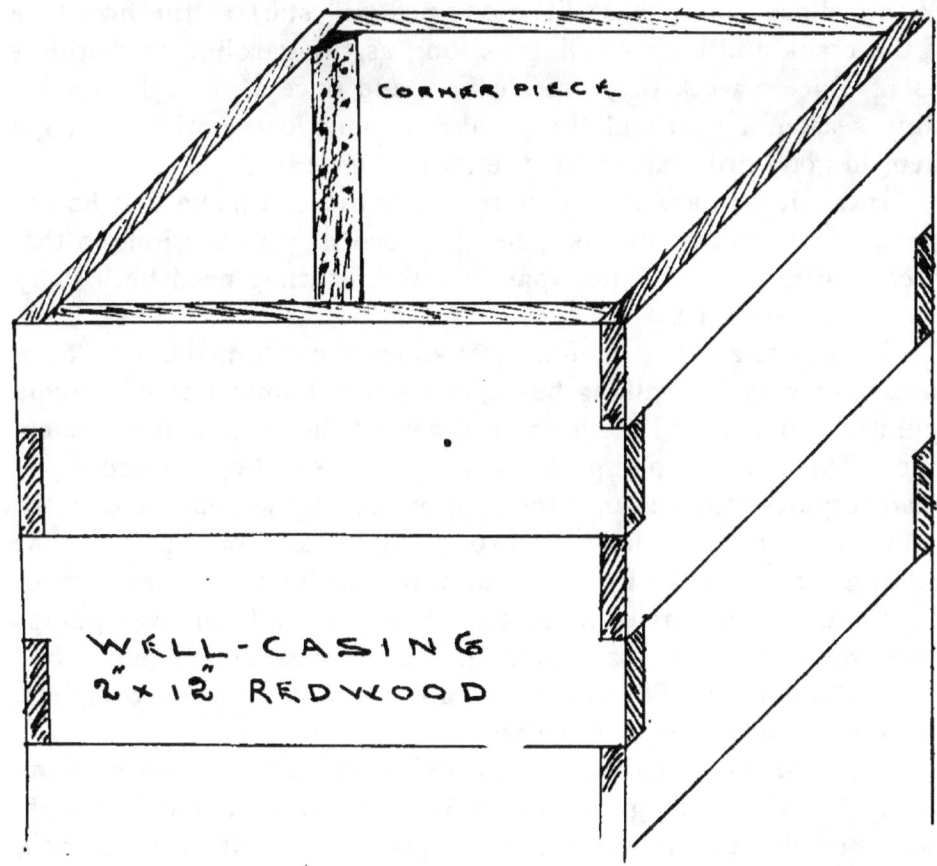

FIG. 32

The well when finished cost as follows:

Digging, 30 feet @ $3 per foot...........$ 90
Digging, 48 feet @ $4 per foot............. 192
Casing lumber 75
Timbering 10
Spikes 3
_____
$370

The engine house is built over the well, with 10x14 feet floor space, and has a good light trap in the floor to prevent rats, etc., from tumbling into the water. This house is very plain, just a frame, and cost $100.

I looked about a good deal before deciding on an engine and pump. I finally decided on a 3-horse power "Lambert" gasoline engine, and have been so well satisfied with it, that lately I have installed another and smaller one of the same make at another well.

With ordinary care, these little engines will start at the first turn of the crank and keep running as long as the gasoline or distillate lasts. Once a week they are cleaned, and once a year, during the rainy season, I overhaul them, take up any lost motion, and get them in good order again for the irrigating season.

In buying an engine it is wise policy to get a make that has an agency near by, and that is generally used in your section, so that in case of repairs, or a new part wanted, no time need be lost by having to send east.

Instead of a patent pump, I got an engineer to make me a common, old fashioned walking beam, and put a double action 4″ pump cylinder at one end of the beam and geared the other end to the engine. This style of pump and gearing is easily kept in order. A small trap over the pump in the roof of the engine house allows the pump and rod to be hoisted through it in case of repairs. The walking beam means a good deal of power with a very small engine. My engine never carries more than half her load, and yet pumps three thousand gallons an hour with a consumption of six cents' worth of distillate. The engine, pump and walking beam complete, set up and ready to run, cost $340.

A six thousand gallon galvanized iron tank, set on a frame twenty feet above the ground, cost $120. The tank is ten feet high, and when full gives fifteen pounds pressure to the square inch, falling to ten pounds when nearly empty. I find this pressure sufficient for all requirements.

A tank made of No. 18 galvanized iron will last twenty years or more with proper care, and is, in my opinion, preferable to wood for many reasons.

The usual practice is to pitch the inside of the tank, but I find this a mistake; the tank is distended when full and contracted when empty, and these motions tend to crack off the pitch and leave bare iron, which, in turn, rusts, although galvanized; so that in a very few years leaks occur.

The best plan is to have a "union" close to the tank on the inlet pipe, and another on the outlet pipe, and after getting the tank in its place on top of the tower, block it up a couple of feet and give both inside and outside two coats of red lead, allowing a week or ten days between each coat. Personally see that the paint fills every crack and corner, especially around the rivet heads, and joints of the plates. When thoroughly hard, lower the tank and connect the piping.

The "unions" will make this an easy thing to do every year; but only one coat will be necessary after the first time; and if a warm day is selected the tank can be emptied, cleaned, dried and painted in one day, and filled again the next morning.

In getting piping, always get "dipped" pipe; that is, pipe that has been dipped in hot tar. It will cost from half to one cent a foot more than black pipe, but the tar preserves the iron, and the pipe will last a great deal longer and be less liable to choke up with rust. A glance at the "general chart" of the place shows the system of piping that I use, and, as far as size of pipes is concerned, applies only to where the water comes from your own tank, which means

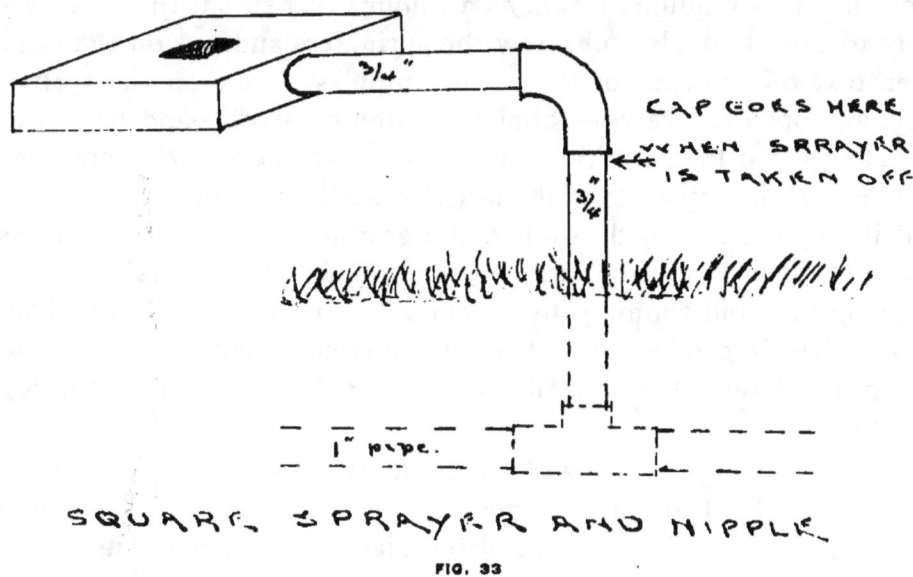

SQUARE SPRAYER AND NIPPLE.

FIG. 33

a limited or low pressure; but where it comes from an outside water service the pressure should be great enough for the use of three-quarter inch pipe all over the place.

A good deal of the water pressure is lost by the friction of the water running through the pipes; the smaller the pipe, the greater the friction. To obviate this as much as possible, I use 1½" pipe from the tank to the lawn where the greatest pressure is required for sprinkling; and from this 1½" pipe I use parallel lengths of 1" pipe, about eighteen feet apart. These 1" pipes go through the lawn, and have a three-quarter inch nipple about a foot long every eighteen feet. Each nipple has a cap, which when I sprinkle, is taken off and a square sprinkler screwed in its place (Fig. 33). I generally have three sprinklers going at once, each spray overlapping the other. I find this a much cheaper and better way than having two or three hydrants on the lawn and using a rubber hose, that costs from fifteen to twenty cents a foot, and has to be carted around every time the sprinklers require shifting. Besides, a lawn sprinkled with hose and a circular sprinkler has to have a good deal of hand sprinkling done, where the sprinklers have not reached. By the method that I use, one has only to screw on the three square sprinklers in a line, and open the valves. At the end of an hour a strip of lawn, about eighteen feet wide and fifty-four feet long, will have been thoroughly and evenly sprayed with about fifteen or eighteen hundred gallons of water. I find that each sprinkler uses from five to six hundred gallons an hour. When shifting, one has only to close the valve, unscrew the sprinklers and put on the caps, then take off the caps of three more nipples, screw on the sprinklers, and open the valve—about five minutes' work—and no worry about seeing if the sprinklers are in the right place. Of course, the nipples sticking up all over the lawn do not improve the looks of it; but if they are nine inches under the ground and only three inches above it, they do not look so very bad. Care must be taken when mowing to avoid running into them with the mower. If one happens to have a careless man it is easy to stick a stake in the ground alongside of each nipple until the mowing is finished; this will jog his memory.

I use one-inch pipe, with three-quarter inch hydrants, in the orchard and kitchen garden, where a good deal of water is required for irrigating; but for the rest of the plant, three-quarter inch pipe is all that is required.

The General Chart shows a good deal of piping and a great many faucets; but one must remember that there is no time to waste in carrying water. In buying the pipe, watch a chance for a low market; the usual price is about five cents a foot for three-quarter inch dipped pipe, eight for one inch, and twelve for inch and a half, but the market varies a good deal, sometimes going below these figures, and again, rising above them. The faucets or bibbs are about sixty cents each when bought by the dozen. Get your own tools for fitting and laying the pipe; a set consisting of pipe-vice, pipe-cutter, two pipe tongs or pipe wrenches, and a box of dies, will cost from $25 to $30, and by doing your own work quite a saving will be effected. The work is pleasant, and very simple to learn; in fact it is necessary to know how to fit piping and to have a few tools, for a broken or leaky pipe is a common occurrence where much piping is used. When connecting pipes, use graphite mixed with oil, instead of red lead; it is much easier for disconnecting, as red lead gets very hard.

I use a square sprayer called the "California Square Sprinkler," it is the only square one that I know of, and is, I believe, no longer made. I am making one on a plan of my own, and if it proves better than the ones I am using, I will put it on the market.

CHAPTER XV.

ORCHARD, GARDEN, LAWN, ETC.

In each hen yard are planted a couple of pepper trees, which grow fast and afford a fine shelter both for sun and rain. I have some four years old that are twenty feet from side to side and quite twenty feet high. For the first year or two they will require a few buckets of water during each of the hot months; after that, the roots will be deep enough to do without irrigation.

Gophers are fond of young peppers, and soon kill them by eating the root. I circumvent them by buying the little trees in quart tins, and then getting some empty five gallon oil cans, I knock out the bottoms and lay two narrow pieces of shake where the bottom was, fill up with soil, take the peppers out of the quart tins, and plant them in the oil tins. I then dig my hole and plant the tin, keeping an inch of it above the ground. By the time the tin has rusted to pieces the roots will be big enough to withstand the gophers. Before transplanting the peppers from the small to the large tins, water the soil well, and then cut the can down the side and spread it open; the plant can then be taken out without disturbing the roots.

Some people will naturally say, "Why not plant fruit trees in the yards instead of peppers?" Well, there are two reasons. One is, that fowls make the trees and fruit very dirty. But a more important one is that fruit trees, laden with ripe fruit, are a strong temptation to boys, who might easily cut the fence at night to get at the fruit. The fowls are not so easily got at; they are locked in, and besides, are liable to make enough noise, when tampered with, to arouse a self-respecting watch dog.

I have a small home orchard which cost me a good deal to plant; but the way the trees have grown more than repays me for all the trouble and expense. The soil is only about twelve inches deep; under that is seven feet of hardpan, and under that, water gravel. I knew that if I planted on top of the hardpan, the roots, being unable to penetrate it, would spread through the shallow soil and require constant irrigation, and, at the end of a few years, peg out for want of nourishment.

I therefore dug a hole for each tree, from six to eight feet in diameter, and four or five feet deep; from the bottom of this a hole was drilled through the remaining hard pan into the gravel, and a couple of sticks of giant powder tamped into the drill holes and exploded, thus breaking up the hardpan into small pieces. The hole was then tested by running water into it, and if it drained out quickly, I knew that it was all right. A wheelbarrow load of sand was then dumped into the hole and spread over the bottom to prevent the hardpan from caking again, and the hole filled up with good soil.

When finished, these holes had cost me about a dollar and a half each, so, naturally, I spared no expense to get first class trees, true to name, to put into them. Mr. Luther Burbank kindly advised me in selecting some of the varieties of trees likely to do well in my locality, and the balance I selected myself through Professor Wickson's "California Fruits." This book, and "Vegetables," by the same author, have proved of the greatest service to me. I raise nearly all the fruit the family requires, and have a greater variety than I can get in the local market; no bought fruit has, for me, quite the same taste as the fruit I raise and pick myself. I irrigate four times during the summer, but hope that in another year or two the tap roots of my trees will have grown down into the water gravel, and so be able to help themselves to what they need, and save me the necessity of irrigating.

I find the following fruits do well in this locality, oranges, lemons, grape fruit, loquats (large variety), peaches nectarines, apricots, apples, pears, figs, plums, and persimmons. I have several varieties of each of these fruits. There is also a small vineyard, and a berry patch, with blackberries, raspberries, Logan berries and guavas, all of which have done well.

I must not forget to mention the kitchen garden, which gives us all the vegetables we can use, and more. It is no fancy of mind to say that our vegetables are better than anything we can buy, for every one knows that stale vegetables can not possibly compare with those gathered just before using. A dozen beds fifty feet long by four wide, will, with very little labor keep the average family well supplied all through the year. I find it a good plan to have a cold frame, covered with cheese cloth, for sowing the seeds of plants that can be transplanted, as it protects the young plants from their enemies the birds, and from the too friendly sun.

The best way to plant the beds is as follows: After the ground has been dug and raked, sow the seed and then cover the whole bed with a layer of well rotted stable manure, an inch deep (chicken manure is too strong). This can be sprayed with a hose a couple of times a week, as the manure mulch keeps the ground from drying out and caking; in fact, the ground will keep in perfect order until the end of the crop, even through the hottest weather. No hoeing will be necessary, and the weeds are better pulled out than hoed. When the crop is finished, the mulch can be dug in and the new crop planted without digging in fresh manure. Follow a root crop, such as carrots or onions, with a leaf crop, as rotation of crops is good for the soil, and gives the best results. Keep a compost heap of stable manure ripening all the time for the kitchen garden. For the orchard I use chicken manure once a year.

I irrigate the berries by making a large basin around each bush at the end of the spring rains and filling it with mulch; when they are watered these basins are filled and left. This plan conserves the moisture and does away with hoeing. Give them plenty of water during the growing and bearing season.

In starting the lawn, sow it to grain, such as oats or barley, with the first rain, and before the crop ripens, cut it, to prevent any grains from falling and coming up later. In March prepare the ground and get it as level as possible; then sow to white clover, rake in the seed, and give the ground a thorough rolling. It can then be sprinkled with screened stable manure and rolled again.

The object of first sowing a grain crop is to firm the ground, and to kill out the weeds. If there is no rain for a few weeks after sowing the seed, be sure to turn on the sprayers before the ground dries out too much. The screened manure will act as a mulch and retain the moisture to a certain extent; but do not trust too much to it, especially when you have your irrigating system in readiness.

The principal thing in lawn making is getting the ground level and firm, so that the lawn mower will cut the clover evenly, and leave a smooth surface. Well firmed soil is most necessary, too, for the sprouting of the seed, for loose soil does not give it the needful moisture. I believe that half the trouble of seed not coming up in garden and lawn can be laid to this cause.

I am aware that this chapter does not strictly apply to the working of the "Practical Poultry Plant," and will be an old story to some of my readers; but there may be others who, like myself, come from very different occupations and physical conditions, and enter a new field, feeling in the dark as to the best method of procedure. Six years ago I would have been glad of these suggestions, and the memory of my own lack of knowledge at the start leads me to offer them to others who are equally inexperienced.

CHAPTER XVI.

FANCY STOCK

The profits that I have mentioned so far, are entirely from market eggs. The birds that lay these eggs are all thoroughbreds, but are unmated, and the eggs unfertile and sold at market prices. I particularly want to impress this point, that for the first few years a beginner had better look to the egg market for his principal source of income, and then, if he cares to, gradually work into the fancy business, which means raising standard stock and eggs for setting, at fancy prices. It also means steady advertising, and exhibiting and winning prizes at the poultry shows.

After having decided on the particular breed or breeds that you are going to raise, eggs or stock must be bought from one of the best known breeders in the country, at pretty stiff prices. I bought eggs at $5 a setting, and if I got one or two first class birds from each setting, I was satisfied. In this way I have now some fine birds, and each year sell more stock and settings.

I do not exhibit, and do not intend to, as I find nearly all my time is required on my place; and my principal vocation at present is making a success of commercial poultry and market eggs, the fancy branch being entirely a side issue.

An amusing incident happened a few years ago in the town near which I live. A stock and poultry show was held there, and afterwards I was asked why I had not exhibited. On replying that I had been too busy either to exhibit or to attend the show, my questioner, who happened to be the owner of one of the meat markets, replied: "Why, I was the poultry judge, and I am sure that you would have taken some prizes; the birds that I awarded prizes to were much smaller and lighter than your birds, and hadn't much meat on the breast; between you and me, they were poor table birds." I said, "I thought it was an exhibit of fancy poultry." He replied, dubiously, "Yes, I suppose it was, but you see I am more used to judging a bird by its meat than its feathers!"

Please don't accept this yarn as being typical of the ordinary poultry show.

I find that for family use the heavier breeds are preferable in many ways to the light ones. They mature more slowly, and do not get so tough and stringy as the Leghorns. If the cockerels are kept separate from the hens they are good roasters even when a year old, whereas the Leghorns begin to toughen at six months.

The Plymouth Rocks, Light Brahmas and Buff Orpingtons are all good table birds, but the last named breed is by far the choicest, being finer grained and more juicy than the other two. They all lay as well as the Leghorns, with the advantage of being better winter layers. Indeed, even through their moult they lay eggs enough to pay their expenses; but they have to be kept in small flocks of fifteen to twenty-five, to get the best results.

CHAPTER XVII.

THE SEPTIC TANK

One of the serious problems of country life is the disposal of sewage. It bothered me a good deal, up to a few months ago, when a friend gave me a paper from which the following is taken.

I built a tank of brick, first putting in a concrete foundation fully six inches deep, and cemented the tank both outside and inside. I also put in a length of vitrified pipe at the bottom of the first tank, sloping the floor toward it, so that every thing would drain out. This pipe was then plugged and the plug made air tight. When the time for cleaning comes, instead of taking off the roof of the tank and baling it out, I shall draw the plug and poke the hose, with a strong pressure of water, through the plughole. I used for the roof 3"x12" redwood, seasoned and tarred twice before placing, and then made all air tight with a heavy coat of cement. Two or three sacks of common charcoal in the filter is better than sand or gravel. The tank cost me exactly $100, and I consider it the best investment I ever made.

Disposal of Sewage on the Farm.

At many country homes where it is desired to introduce modern improvements in the way of waterworks to supply bath room, closet, sink and laundry, the disposal of sewage is quite a serious problem.

Fortunately it is a problem quite easy of solution by the "Septic Tank" system at once scientific and simple though but little known. The system can be easily applied in any place where sufficient fall can be secured to carry away the sewage. It is inexpensive, absolutely automatic and thoroughly effective and satisfactory. It can perhaps be best illustrated by describing a plant now in operation at the Western Hospital for the insane at Watertown, Ill. The system is the result of an accidental discovery, and was first put in successful practice by Dr. W. E. Taylor, superintendent of the above named institution though now being installed at other public institutions in Illinois and attracting much attention elsewhere. That it is perfect in its action may be gathered from the fact that it receives all the concentrated sewage from an institution whose inmates and employes number nearly eight hundred people, thoroughly and completely disposes of all organic, effete and poisonous matter with no residuum or deposit, and the product flows away in a clear, sparkling stream of water, ninety-eight per cent pure by chemical analysis when it strikes the air, the remaining two per cent of impurities being liberated upon exposure to the atmosphere, leaving a stream of clear water pure enough for any purpose whatever. That this sewage can enter at one end of a tank a foul, offensive stream reeking with filth, and emerge from the other end a limpid stream of water actually pure enough to drink, seems wholly incredible, and yet such is the case and the wonder of it all is that it cleanses itself automatically, without any artificial agency, solely through the work of the filth bacteria preying upon each other. This system works continually, summer and winter, year in and year out, with absolutely no attention and without change in any season, never freezing. It is practically adapted to use in the country at a distance from city sewers, and even for the use of towns and cities is entirely reliable and effective.

At the Watertown Asylum the system consists of two oblong tanks of seventy thousand gallons capacity each, placed side by side, one tank emptying into the other through a pipe. For all practical purposes, however, one tank with a weir box at one end, is exactly as good as two tanks, as it has been found that the water as it emerges from the first tank is just as pure as after it has passed through the second tank. The object of this weir box is to check the overflow and prevent any agitation of the sewage in the tank.

The tanks in this system are located about a quarter of a mile from the buildings. They might be located forty feet or four miles away, according to convenience, the result would be the same.

The sewage tank as shown in the illustration, consists of a brick box with eight-inch walls and floor, lined within and without with cement. Concrete would make a better tank. The roof is made air tight with a heavy coating of pitch and all crevices are tightly sealed with the same material. The sewer inlet is about two feet below the surface of the sewage in the tank. A short distance from the opposite end of the tank a cross wall is built, having a narrow opening extending across the tank on a level with the inlet. This opening has little if any greater capacity than the inlet. Such an opening causes less current in discharging than would a circular opening. In the end wall is a row of curved tile so placed that the outlets are two feet above the sewer inlet and the opening in the cross wall. The cross wall forms a weir, or dam, which retards the outflow from the main tank, and of course there can be no discharge until the contents of the tank and weir box reach the level of the curved tile outlets. Thus both inlet and outlet are submerged about two feet below the surface of the sewage in the tank. The filter box is filled with sand and gravel and has an outlet at the bottom through which the water finally discharges.

The operation of this system is simplicity simplified. The sewage entering the tank remains until it fills the tank and the weir box to a level with the overflow from the curved tile outlets. In twenty-four hours or a little over, after entering the tank a scum will have formed on the surface, an inch or more in thickness, consisting of a solid mass of filth bacteria, which prey upon the poisonous matter and the solids contained in the sewage, constantly fighting among themselves and destroying each other like the Kilkenny

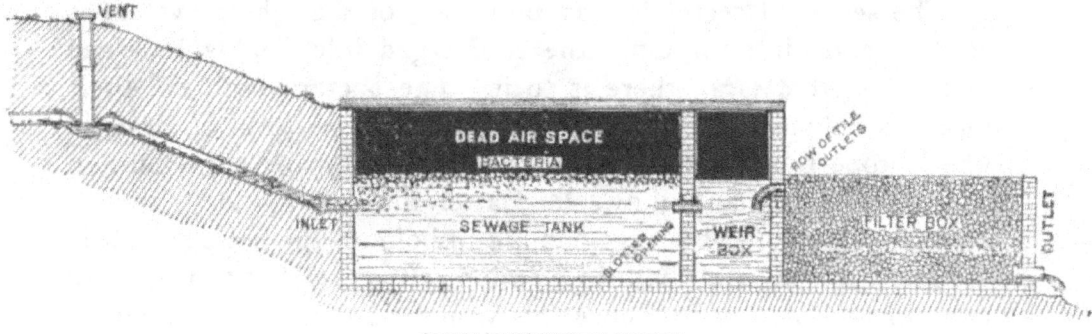

SEPTIC TANK SEWAGE SYSTEM.

cats, which devoured each other until nothing was left but the tail, the tail in this case being represented by the two per cent of poisonous matter left in the water as it escapes, and which is at once eliminated upon exposure to the atmosphere.

Light and air are fatal to these bacteria, hence the necessity of keeping them in a dark, air tight place that they may accomplish their work. For this reason the tank must be air tight. Again to do their work effectively they must be left in perfect quiet, hence the inlet and outlet are submerged below the surface in order that from inflow and outflow as little current as possible may be caused, and this quiet is further assured by means of the weir box.

Upon emerging from the tiles the water is clear as crystal, and by chemical analysis contains but two per cent of bacteria that would be in the slightest degree injurious to the human system.

This water is allowed to filter through the sand and gravel, its exposure in this manner to the air destroying all remaining bacteria, so that it emerges from the final outlet absolutely pure.

Knowing its source, one would not care to drink it, though it is pure enough for this purpose, and stock may drink it with perfect safety.

A system of this kind will not freeze in winter, as the gases arising from the sewage in the tank generate enough heat to counteract the cold and prevent freezing. The water as it emerges will be found much warmer than the air in cold weather.

In cases where the sewage discharge is scanty and intermittent there might be danger of the water freezing in the filter box during a long cold spell, and then it would be advisable to erect a small tight building, well protected from frost, over the whole outfit, including both tank and filter, but when the sewer is in constant use this would be unnecessary.

The secret, if secret it may be called, of the whole system is the dark and air tight tank, the submerged inlet and submerged outlet, and that is all there is to it. The bacteria will do their work if let alone. If stirred up they refuse to perform as desired. When properly working the tank might be opened, the bottom scraped and not a handful of solid matter could be found.

The tank should be large enough to hold all the sewage that is ever likely to run into it within a period of twenty-four to thirty-six hours. For a private residence this would rarely need to be larger than three feet wide, six feet deep and eight to ten feet long.

SUMMARY OF EXPENSES

Land, 5 acres @ $200 per acre..........................$1000
Store House 230
No. 1 Brood House 200
No. 2 Brood House, containing incubator and feed rooms... 240
12 inside sections of Brood House....................... 12
12 Brood House yards 70
2 Cyphers' 360-egg Incubators 80
12 Brooders 120
12 Chicken Houses 120
12 Chicken Yards and fixtures 133
12 Hen Houses, complete.............................. 780
12 Double Hen Yards and fixtures....................... 430
2 Compost Houses, with sand boxes..................... 10
2500 feet ¾-inch Dipped Pipe @ 5c per foot.............. 125
2 dozen Faucets, or bibbs 14
Set of Piping Tools................................... 25
2 Post-Hole Diggers and Wire Stretchers................. 10
Carpenter's Tools, etc................................ 51
Labor, 6 months @ $50 per month...................... 300
Stock, or Eggs, Feed, etc............................. 550

Complete Plant $4500

A YARD OF COCKERELS

GATHERING EGGS

A YARD OF PLYMOUTH ROCKS

BRAHMAS MOULTING

www.ingramcontent.com/pod-product-compliance
Lightning Source LLC
Chambersburg PA
CBHW08083822O526
45467CB00008B/2317